To Cheryl –

Bonnie Bladgett [signature]
11/12/09

Jewel OF COMO

The Marjorie McNeely Conservatory

Leigh Roethke *and* Bonnie Blodgett

AFTON PRESS

Dedicated to the memory of

Donald McNeely
(1914–2009)

and

Marjorie Reeds Brooks McNeely
(1916–1998)

who raised us with an appreciation for the natural world

whose legacy is nowhere more evident than in the

MARJORIE MCNEELY CONSERVATORY

Robin Brooks Tost
Peter Kendall Brooks
Greg McNeely
Kevin McNeely
Nora McNeely Hurley

The publication of

Jewel OF COMO

THE MARJORIE MCNEELY CONSERVATORY

is made possible by major gifts from the

DONALD MCNEELY FAMILY

H. G. MCNEELY, JR.

and

COMO FRIENDS

with additional generous contributions from

Katherine B. Andersen Fund

Mary Anne Anderson

F. R. Bigelow Foundation

Harry Drake

Elizabeth and John Driscoll

Joan Duddingston

Alexandra and Robert Klas

Malcolm and Patricia McDonald

Mike and Shirley Miles

Ford and Catherine Nicholson

Constance Otis

Kari and Dan Rominski

Front cover: The Marjorie McNeely Conservatory.
Photograph by Tina Dombrowski.

Back cover: The Sunken Garden, winter flower show.
Photograph by Tina Dombrowski.

Copyright © 2009 by Afton Historical Society Press
ALL RIGHTS RESERVED
First edition

No part of this book may be reproduced or transmitted in any form or by any means, electronic or mechanical, including photocopying, recording, or by any information storage and retrieval system, without permission in writing from the publisher.

Designed by Mary Susan Oleson
Production assistance by Beth Williams
Printed by Pettit Network Inc., Afton, Minnesota

Library of Congress Cataloging-in-Publication Data

Roethke, Leigh.
Jewel of Como:the Marjorie McNeely Conservatory/
by Leigh Roethke and Bonnie Blodgett.—1st ed.
   p. cm.
ISBN 978-1-890434-79-3 (hardcover:alk. paper)
1. Marjorie McNeely Conservatory (Saint Paul, Minn.)
2. Parks—Minnesota—Saint Paul—History.
I. Blodgett, Bonnie. II. Title.
QK73.U62M37 2008
580.73'776581—dc22

2008002498

Printed in China

Patricia Condon McDonald
PUBLISHER

AFTON PRESS
P.O. Box 100, Afton, MN 55001
651-436-8443
aftonpress@aftonpress.com
www.aftonpress.com

Table of Contents

FOREWORD 10

CHAPTER ONE 13
a place to Dream

CHAPTER TWO 29
Land Grab

CHAPTER THREE 41
the Tropicals under Glass

CHAPTER FOUR 59
Jewel in the Crown

CHAPTER FIVE 69
Show time!

CHAPTER SIX 85
another Rescue

CHAPTER SEVEN 103
the Gift

ILLUSTRATION CREDITS 122

INDEX 123

Foreword

JEWEL OF COMO is a fascinating story—from the first purchase of land in 1873 by the City of St. Paul and its early development, to the completion of the conservatory in 1915, its close calls with ruin, and its just-in-time rescues. A place for contemplation, relaxation, inspiration, and enjoyment of nature's beauty, the Marjorie McNeely Conservatory is also a wonderful example of public and private partnership. Most recently, this collaboration culminated

in a generous gift from the late Don McNeely that honored his wife, Marjorie, and provided for the glorious, present-day renovation of and additions to the conservatory.

Many kids today experience "nature deficit" as they spend much more time indoors with technology than they do outdoors. Here is a wonderful opportunity for young people to learn about and enjoy the world of plants! In the future, as society becomes more engrossed with technology, the conservatory and its counterparts worldwide will become ever more important to help us connect with the wonders of the natural world.

Plants are the source of all of our food. Plants harness the energy of the sun through the magic of photosynthesis and, indeed, all of our food ultimately comes from plants. One can see in the conservatory many fruiting plants that do not grow in Minnesota, including bananas, figs, cocoa, and citrus.

Five generations of our family have enjoyed visiting Como Park and the Marjorie McNeely Conservatory. We visited the conservatory on St. Patrick's Day this year to seek inspiration for this writing. As we watched the many children delighting in the sights and sounds and smells, we recalled our happy visits as children and later with our own children and grandchildren. Many of our visits have been in winter months when there is very little green outdoors in Minnesota. Gordie describes these winter trips to the conservatory as "the poor man's trip to Florida."

We always marvel at the numbers of people visiting the conservatory, and their diversity: families with small children, teenagers and the elderly, people of many racial and economic backgrounds. The Marjorie McNeely Conservatory is a remarkable melting pot that can be enjoyed free of charge by people from St. Paul and Minneapolis and beyond. A testament to how much people treasure the free zoo and conservatory, the donation boxes near the entrance to the Visitor Center each year collect over a million dollars.

Enjoy this story of St. Paul's splendid glass house at Como Park—a jewel for public enjoyment and learning—then plan your visit to the Marjorie McNeely Conservatory at the earliest opportunity!

Gordie and Jo Bailey
BAILEY NURSERIES, NEWPORT, MINNESOTA

The Schiffman Fountain is located south of the lakeside pavilion.

CHAPTER ONE

a place to Dream

PEOPLE STILL SPECULATE on why the small body of water everyone used to call Sandy Lake, on account of its sandy bottom, started showing up as Lake Como on area maps in the mid 1800s. Some trace the name change to an immigrant potato farmer who owned land along its shores and often remarked that the lake reminded him of the original Lake Como in the Western Alps, where he'd spent his boyhood. Others credit Henry "Broad Acres" McKenty, a flamboyant land speculator. McKenty arrived in St. Paul in 1853 with money enough to purchase most of the land surrounding Sandy Lake, including that potato farm, and visions of a lavish resort district dancing in his head.

Minnesota in 1853 was a far cry from Italy. Most of the territory was still forest and prairie, and its future capital, St. Paul, was little more than a rugged hamlet occupying about ninety acres at the upper terminus of the Mississippi River steamboat trade. The town proper consisted of a few wood cabins and tiny shops, mercantile firms, a church, six hotels, and a saloon. Alexander Ramsey, Minnesota's territorial governor, declared his adopted home "emphatically new and wild."

Stereographic view of Henry McKenty's residence, upper left, at Lake Como.

Statehood, granted in 1858, quickened the pace of progress. Newcomers set to work buying land, erecting buildings, charting railroads, and building roads. Impulsive to a fault, "Broad Acres" McKenty put up $6,000 of his own money to build a road connecting the St. Paul residents to Lake Como. While he failed to secure titles for all the land the road covered, few people objected to progress in those days.

Restauranteur Otto Adler had a resort hotel up and running by 1857 and built himself a fine, red brick manor house to go with it. Two more hotels and a handful of summer homes followed, including McKenty's private residence. McKenty stocked the lake and rented out fishing boats. Some began calling Lake Como by *his* name instead, even the newspaper reporter who, in 1860, wrote about the lucky angler who pulled a twenty-pound pike out of "McKenty's Lake."

Lake Como had something for everyone: rowing regattas, shooting matches, soirees, and horseback rides for those who could afford them; the Aldrich Hotel bowling alley and merry-go-round for those who liked rowdier pursuits. By 1862, the St. Paul and Pacific

Sailing was one of many activities offered at the Aldrich Hotel on Lake Como, pictured here about 1870.

Line (later the Great Northern Railway) had laid track just south of Lake Como. The following summer, an omnibus began conveying people back and forth from town, charging fifty cents round-trip. But that was not the only way to get to Lake Como; a livery stand on Como Road that competed for fares gained notoriety when its drivers began racing their rigs to pass the time on slow business days.

By this time, St. Paul's transformation from wilderness outpost to bustling transportation, commercial, and government hub was well underway. New arrivals to the state were pouring in; census figures doubled each decade from statehood until after the turn of the twentieth century. Commercial and municipal buildings in the old downtown pushed neighborhoods farther into surrounding farmlands and across the river. As green space near town became scarcer, it also increased in value.

Like most young towns, St. Paul had few public parks. Three small plots of land had been donated for that purpose in 1849 by a group of St. Paul's leading businessmen. Henry M. Rice's land in the heart of town was christened Rice Park. Cornelius S. Whitney and Robert Smith donated Smith Park, at the intersection of Sibley and Fifth Streets. John R. Irvine gave the plot that became Irvine Park in Lowertown. Ramsey County got Courthouse Square, on the east side of town, between Fourth and Fifth Streets, the following year.

But they were parks in name only. The city paid little attention to these patches of greenery. Cows grazed on the grass and women strung ropes between the trees and used them to dry laundry. The parks were left to the cows and the laundry for nearly twenty years until finally, in 1867, the Common Council of the City of St. Paul shook off its indifference to the parks and formed a committee to protect urban green space and even add more.

Still, a parcel of land ten miles northeast of town seemed impossibly remote—even one that contained a pretty lake. City fathers had their hands full providing passable roads and a sewage system for the inhabitants of St. Paul; but in a few short years the city's burgeoning citizenry began to crave a natural landscape to escape to on evenings and weekends. Lake Como beckoned during every season. So in 1873 the city bought up all the land around the lake and hired landscape architect Horace Cleveland from Lancaster, Massachusetts, to design a public park like New York City's Central Park. Such expansive, naturalistic oases were called pleasure grounds.

Naturalistic, however, doesn't mean untamed. Cleveland brought in thousands of plants, which needed help adapting to the harsh weather. In similar parks across the country and in Europe, the first "hothouses" had been erected to overwinter tender young plants. As formal gardens designed with tropical plants grew in popularity, more glass enclosures were built for palm, banana, and fig trees, exotic water lilies, and flowering annuals that had to be grown from seed every year, sometimes as early as February or March.

Greenhouses made spectacular floral displays possible, and park developers began to see such displays—and eventually the greenhouses themselves—as major attractions. By the turn of the twentieth century, simple cold frames, in lean-to style, had morphed into full-blown conservatories, architectural marvels that were at once a throwback to medieval Gothic cathedrals with their huge stained-glass windows supported by flying buttresses and a harbinger of modern

Joseph Paxton with Decimus Burton designed the Great Conservatory or Stove at Chatsworth for the Duke of Devonshire in 1837. A huge cast-iron heated glasshouse, it was the largest glass building in the world and the forerunner of the modern greenhouse.

glass skyscrapers. Indeed, Joseph Paxton—creator of the first expansive glass house in England—is regarded as a seminal figure in the history of architecture. His innovative use of glass spurred manufacturers to produce stronger and more flexible building materials. His designs inspired myriad lookalikes, from the tiny terrariums that sat on Victorian nightstands to giant enclosures spanning several acres.

At numerous world's fair exhibitions, including New York's first world's fair in 1853, glass houses pulled in crowds. American public-park planners, too, were intrigued by the new "crystal palaces," as they

a place to Dream

The remarkable domed, iron-and-glass crystal palace built in Bryant Park for New York's first world's fair in 1853.

Jewel of Como

were sometimes called. Such structures offered year-round access to tropical splendors that were sure to lift the spirits of city residents. Also, and perhaps more important, these buildings seemed to prove by their very existence that *this* city wasn't just another boom-and-bust blip on the landscape. Any place with a crystal palace must also have real amenities, culture, and a bright future.

This book tells the story of how the "jewel of Como Park" came to be, and how one of the most modern and sophisticated glass houses on earth when it was completed in 1915 survived the changing tastes and fashions of the twentieth century. What combination of miracles coalesced to save St. Paul's conservatory from extinction over the next century, as so many buildings like it in other cities and countries fell into disrepair and were then sacrificed to the wrecking ball?

Now called the Marjorie McNeely Conservatory in honor of a woman who loved the building all her life, Como Park's own crystal palace stands as a monument to the extraordinary ingenuity, tenacity, and pride of thousands of individuals, many of them volunteer workers. These were people who not only saw St. Paul as a city bound for glory but who also believed that its hardy and hard-working citizens deserved a tropical paradise to warm their hearts in winter and summer, a place to learn about and to wonder at living things, and, above all, a place that would inspire big thinkers like "Broad Acres" McKenty.

A place to Dream

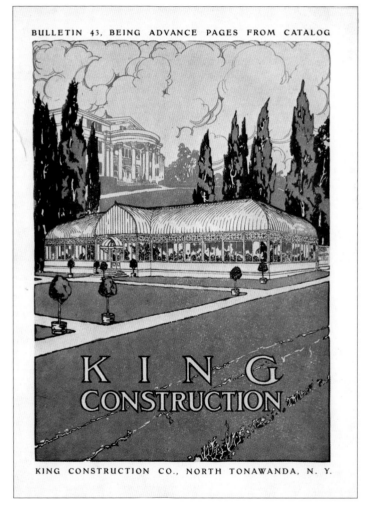

The cover of a 1915 King Construction Company catalog illustrates the lofty ideals of the greenhouse builders.

ART *in the* PARK

IN THE BEGINNING, thirsty park visitors dipped tin cups into the waters of the spectacular Sterk fountain and the stagy "grotto" fountain. Then came the more hygienic Commercial Club and Shipley fountains. In 1905, the park's neoclassical Mannheimer Memorial Fountain was designed by Cass Gilbert (also the architect of the Minnesota State Capitol). Its magnificent stairway, edged with white Italian marble, led up to a pergola whose four Doric columns supported a wooden trellis with a claw-footed pedestal basin residing in splendor below.

A dapper bicyclist stops to cool off at the water grotto in 1895.

A family climbs the steps of Cass Gilbert's Mannheimer Memorial in Como Park.

Dr. Rudolph Schiffman, one of the original park board members gave the park its next fountain, this one a mermaid holding a shell in her hands that spouted water, encircled by dolphins. Schiffman had seen one like it while traveling in Spain and decided that St. Paul, too, should have a bronze siren. The mermaid was placed in a retaining basin at Como Park's Circle Drive, testimony to the park's pledge that gifts be "in accordance with the highest standards of art."

But Horace Cleveland's Como Park was supposed to have been a work of art itself, with any other form of artistic expression strictly prohibited. So much for landscape design dogma. The American Park and Outdoor Art Association, a national organization of landscape architects, artists, and community leaders formed in the late 1890s in opposition to the City Beautiful Movement, tried to set high standards for public art. Its mission statement included these words:

The Schiller Monument in Como Park.

There should be no place . . . for granite pantalooned remembrances of dead musicians and soldiers and statesmen. If we cannot teach people to realize that they should keep their effigies of statesmen where they belong, then let us hide them in thickets . . . We should put nothing in our parks which suggests unrest or anything disagreeable, or that will frighten children, but we should put in objects that will suggest woods, trees, water, and nature.

This fell on deaf ears. In 1907, the U.S. German Societies of St. Paul added a life-sized bronze statue of German philosopher and poet Christoph Friedrich von Schiller, whose positions on human rights and political freedom resonated deeply with Americans of German descent. Not to be outdone, a Norwegian fraternal organization presented a bronze bust of Norwegian playwright Henrik Ibsen to the city. It was placed in Como Park on an eight-foot granite pedestal. Jacob Fjelde created the bust from a life mask of his countryman's face that he had cast in Norway in 1885. (Fjelde also sculpted the *Minerva* statue for the Minneapolis Public Library, *Hiawatha and Minnehaha* in Minnehaha Park, and a bronze statue of Norwegian violinist Ole Bull for Loring Park in

Minneapolis.) Ibsen overlooked Lexington Avenue in Como Park for three quarters of a century before he was plucked from his pedestal and carried off into the night. The whereabouts of the bust remained unknown until 1993, when it was found in a Robbinsdale video store wearing a t-shirt and baseball cap. The bust was promptly restored to its pedestal in Como Park.

A more modern and arguably artistic style of statuary began to appear at Como in the 1920s and '30s. Frederick Crosby's 1927 gift of a granite frog from Japan was a delightful addition to the lily pond adjacent to the conservatory. So were the *Excedra* overlooking the pond, the replica of *Mercury* (a renowned bronze sculpture by Flemish-Italian Renaissance artist Giovanni da Bologna), and the cast-pewter *Aphrodite* greeting garden visitors who arrived through a Doric-columned pergola and up two steps flanked by massive Roman vases decorated with reproductions of Danish sculptor Bertel Thorvaldsen's allegorical *Night and Morning*.

In 1932, then-superintendent George L. Nason designed an armillary sphere for the peony gardens south of the conservatory. "No garden of pretension is complete without a sundial," he said.

Como had moved so far away from Cleveland's

A visitor admires the fountain in Aphrodite's Garden.

vision that when St. Francis of Assisi found sanctuary in the conservatory's North Garden after suffering a blow to the head and the loss of several fingers to vandals in 1957 (it had been located on park grounds), it wasn't the statue's presence in the park that prompted the move indoors, but its safety. Indeed, the only sticking point was whether the statue should be repaired or destroyed. Public support for sculptor Donald Shepard's *St. Francis* (and a $500 grant from the Catholic Daughters of America) settled the matter.

Alonso Hauser's *Reclining Nude #5* cozy amid the succulents.

Park Superintendent George Nason's astrolabe dial, 1937.

In 1962, the Men's Garden Club, acting as trustee for the collection boxes in the conservatory, purchased a fountain and wishing well for the Palm Dome. A plaque affixed to the basin reads: "A coin in the fountain, a wish at this well will bring you happiness, more than you can tell." Placed amid the palms directly under the glass dome, the fountain made a romantic focal point in the center of the conservatory, and was a reliable revenue producer.

In 1965 the fountain was remodeled by former Macalester College art professor Anthony Caponi for use as a base for Como's first work by the talented American sculptor Harriet Whitney Frishmuth, *Crest of the Wave*. A student of Rodin, Frishmuth wrote to the wife of her longtime patron, William R. Anderson, of her wish to create fluid, lyrical works that would give people "a feeling of happiness and uplift." *Crest of the Wave* depicts a young nymph dancing above the water. It was given to the Como Conservatory in Anderson's memory.

The conservatory received its second Frishmuth nymph two years later, after Anderson's family found a copy of *Play Days*, a highly respected work, and had it installed on a fountain basin in a stucco-and-brick niche constructed at the south end of the Sunken Garden.

Harriet Frishmuth's *Crest of the Wave* in the Palm Dome.

A small waterfall behind the statue flowed into the pool and short jets of water shot up around the feet of the female figure. *Play Days* was unveiled during the November 1967 chrysanthemum show.

Ten years later, a Paul Manship statue that had stood in the tiny, triangular Cochran Park in Ramsey Hill since 1926 sparked a long tug-of-war between neighborhood residents and those who wanted the vandalized *Indian Hunter and His Dog* kept in the relatively safe Como Conservatory. Thomas Cochran, the park's

Paul Manship's *Indian Hunter and His Dog* shortly after the dedication of the McKnight Formal Garden in 1967.

namesake, and Manship had been childhood friends. In 1927 Manship had installed the life-size bronze statue in the center of a fountain basin in Cochran Park, adding four bronze Canadian geese at the corners of the basin. It was frequently the target of vandals. Eventually William L. McKnight and his wife, Maude, rescued the statue from further degradation. *Indian Hunter and His Dog* was moved to what became the McKnight Formal Garden in Como Park—but not for long. After Ramsey Hill rose from its squalor in the 1980s, proud neighborhood residents lobbied successfully to have the statue, now worth well over one million dollars, returned to Cochran Park.

The latest addition to the conservatory's art collection is a coin-operated animatronic figure created by Dean Lucker and donated in 2005 by Doe Hauser Stowell in memory of her husband, James Stowell. The statuette depicts a man leaning against a flowering tree while holding up a lighted carousel. With the deposit of a quarter, the carousel spins and the viewer receives a fortune. The whimsical statuette is the most recent expression of St. Paulites' strong connection to Como Park and its conservatory, and the desire to enhance it and to share with it their own favorite works of art.

Harriet Frishmuth's *Play Days* in the Sunken Garden.

Wildflowers edge a path along Lake Como.

CHAPTER TWO

land Grab

AMERICA'S PUBLIC PARK movement began in the 1840s. U.S. reformers published treatises warning of a growing disconnect between man and nature. They encouraged cities to supplement their existing public spaces—the town square—with expansive urban parklands.

"Each town should have a park," wrote Henry David Thoreau, "or rather a primitive forest, of five hundred or a thousand acres." Such thinkers held that contaminated air and crowded streets impaired the lungs and spread illness, and that cramped cities undermined basic humanitarian urges. A walk in the park, they argued, would rejuvenate individuals and, ultimately, society as a whole.

Park proponents also appealed to the vanity of wealthy Americans, slyly referencing the more "cultured" societies of Europe. Andrew Jackson Downing, editor of a leading

Andrew Jackson Downing,
America's first landscape architect.

gardening and landscape-design journal, *The Horticulturist*, praised the "true democracy" of new European parks (former private estates that had been opened to the public), where people from all walks of life came together in the common enjoyment of nature. He likened parklands to museums:

> *Open wide, therefore, the doors of your libraries and picture galleries, all ye true republicans! Build halls where knowledge shall be freely diffused among men, and not shut up within the narrow walls of narrower institutions. Plant spacious parks in your cities, and loose their gates as wide as the morning, to the whole people.*

Lacking Europe's geographic heritage—large estates that could be opened to the masses—but still possessing sizable tracts of undeveloped land on the periphery of its cities, American planners knew the time was ripe to design public green spaces into the country's cultural infrastructure. In fact, it was now or never.

In 1851, President Millard Fillmore asked Downing to devise a landscaping scheme for the National Mall in Washington, D.C., then reviled as the "national cow paddock." Washington's National Mall would have been the nation's first such public park—had the designer not perished in the explosion of the Henry Clay before his plan was executed. That distinction went instead to Central Park. Frederick Law Olmsted and an Englishman named Calvert Vaux (Downing's former partner) developed their "Greensward Plan" for a city-sponsored competition in 1857. Of the pair, Olmsted was the chief designer. He prophesized a time when "New York will be built up . . . and the picturesquely varied, rocky formations of the island will have been converted into formations for rows of monotonous straight streets, and piles of erect buildings."

He proposed, and won approval to create, "a vast wilderness oasis in the heart of a teeming metropolis." To make his "oasis" as appealing as possible as the city grew up around it, he applied English landscape-park design principles in the sculpting of meadows, forests, and lakes out of the eight hundred-plus acres. Like Downing, Olmsted felt passionately about the social advantages of public parks. He wrote that the landscape architect's job was to balance the intentional beauty of stone and mortar with "the [natural] beauty of the fields, the meadow, the prairie, of the green pastures, and the still waters.

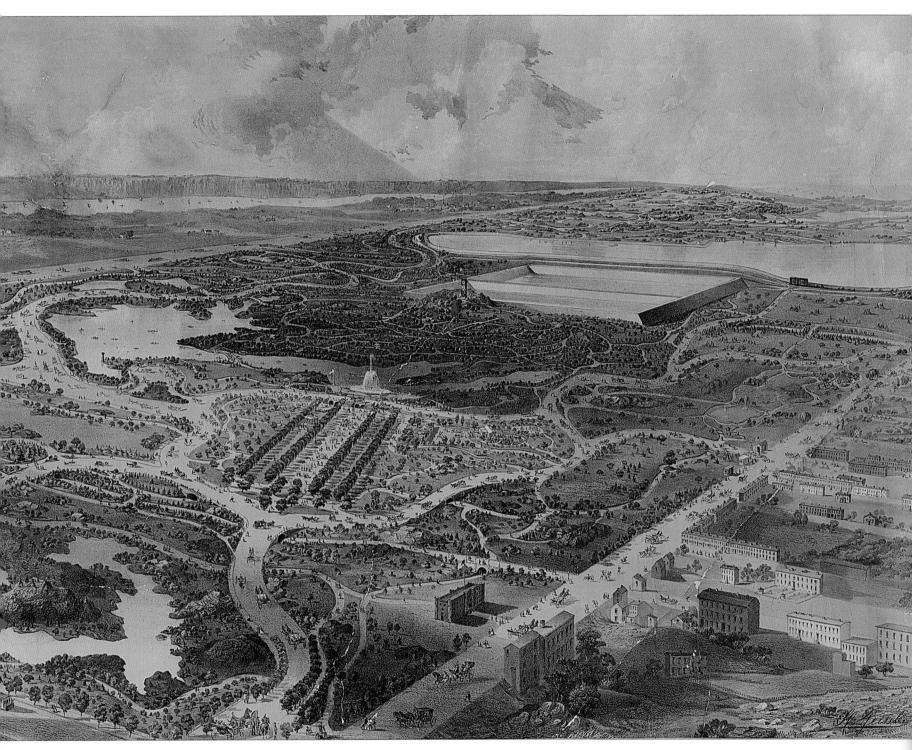
Pierre Martel's bird's-eye view, lithographic image of New York's Central Park in 1864.

Frederick Law Olmsted. Painting by John Singer Sargent, 1895.

What we want to gain is tranquility and rest to the mind."

Olmsted's projects shaped urban park development in the second half of the nineteenth century. Though many people still viewed the wilderness as something to be tamed, settled, farmed, ranched, or mined, the notion of setting aside choice real estate for public parks had begun to resonate in America after the Civil War. Cities from coast to coast embarked on their own park projects, inspired by New York's example. These included San Francisco's Golden Gate Park, Chicago's Lincoln Park, Philadelphia's Fairmount Park system, and Boston's Emerald Necklace. California's Yosemite became the first state park in 1865. Seven years later President Ulysses S. Grant signed into law an act to preserve 2.2 million acres of wilderness for a "public park and pleasuring ground for the benefit and enjoyment of the people." Yellowstone National Park thus became the world's first national park.

While Olmsted and his colleagues were working mainly east of the Mississippi, Horace Cleveland was busily staking out the western states and territories. The same year that land was set aside for Yellowstone Park, William Watts Folwell, the first president of the University of Minnesota, invited Cleveland to speak on the fate of the "vast regions lying undisturbed between the Mississippi and the Pacific."

Cleveland's proposal to preserve the natural characteristics of a landscape impressed his audience. The St. Paul Chamber of Commerce requested an encore reading, and eventually asked Cleveland to prepare a formal proposal for a comprehensive city plan. The city council created a commission to "contract for and purchase not less than five hundred, and not more than six hundred and fifty, acres of land within a convenient distance of the city of Saint Paul, but beyond the present limits thereof" for a public park. Five leading citizens—Joseph A. Wheelock, Samuel Calhoun, William Pitt Murray, James C. Burbank, and Henry H. Sibley—were handed the task.

Horace Cleveland was barely twenty-one when he first set out for the wild west from his Lancaster, Massachusetts birthplace to work as a land surveyor in central Illinois and indulge a yearning to explore a true wilderness. Cleveland later wrote that Minneapolis and St. Paul "could not then have been reached in safety by a white man except by steamboat or with an armed escort." After returning east, he established a successful landscaping practice in Boston, but wanderlust struck him again after

Horace William Schaler Cleveland.

the Civil War ended. He landed in Chicago in 1869, this time to stay.

When Cleveland returned to St. Paul in the early summer of 1872, he spent several days examining the topography of Minnesota's capital. He trudged up its hills and along its creeks, surveying the town from many vantage points, including the rural perimeter. On June 24, Cleveland presented his plan, mapping out a future St. Paul that incorporated green spaces, radial avenues, and wide boulevards inspired by Baron Haussmann's Paris.

Cleveland proposed at least two attractive thoroughfares connecting St. Paul with Minneapolis. Now was not the time for monuments, he said. The young city should develop its natural beauty before adorning it with "jewelry." Cleveland also urged the city to acquire as much potential parkland in and around St. Paul as possible before it was priced out of reach. Large outlying parks connected to the city by avenues rounded out Cleveland's plan. The shores of Lake Phalen and Lake Como topped his list of potential sites, as both offered bodies of water large enough to support public recreation and grounds ideal for "artistic use."

The commissioners selected 257 acres just inside the city limits on the north and west shores of Lake Como. Most of the land was purchased from hotel proprietor and farmer W. B. Aldrich and from former Minnesota governor William R. Marshall.

The purchase of the land riled some citizens. A park paid for with public money? It seemed to some outrageous. Fissures appeared in the city council early in the summer of 1873. The *St. Paul Dispatch* reported on a dissenting resolution arguing

> *. . . that the city has parks in almost every ward . . . accessible to the taxpayers of moderate means to visit without being obliged to support horses and carriages, that this investment is a discrimination in favor of the rich that need no protection, as against the men of moderate means, that are the bone and sinew of the country and require the fostering care and protection of the law makers.*

Another resolution accused the city of secretly buying land without public consent, at considerable expense to taxpayers. Popular opinion sided with opponents of the "land grab." In April of 1874, a petition signed by 148 citizens, many of them from St. Paul's business elite, who had been hit hard by the financial panic of 1873, called for the city to sell the land at cost. Rather than a park that would only advance the "private interests of wild real estate speculators," they pressed instead for "sewerage, elevators, free bridges, good roads, and the suppression of houses of ill fame," contending that the park project was far too ambitious for a city of only thirty thousand residents.

Businessman William A. Banning argued that St. Paul would need no parks for another fifty years, when the city's population, he correctly surmised, would number 200,000. He predicted that the park would cost taxpayers $3 million to build, a shocking sum that he nevertheless significantly underestimated.

The debate in St. Paul was hardly unprecedented. In New York City, arguments over whether, where, and at whose expense a grand public park would be built raged for three years. It took Kansas City, Missouri, considerably longer to finally establish a parks board. And just across the river from St. Paul, the Minneapolis City Council frequently passed on opportunities to acquire first-class land for parks. In 1866, the city council had nixed the purchase of Nicollet Island for the sum of $47,500. Four years later it rejected the asking price of $50,000 for 250 acres surrounding Lake Harriet.

But opposition to Como Park evaporated as quickly as it had appeared. St. Paul Chamber of Commerce member Pennock Pusey reminded his colleagues that their financial worries would pass—they did, in short order—as would the opportunity for the city to secure "sightly and characteristic spots for which our landscape is noted" before private investment would make the undertaking far more costly and difficult.

In 1878, the city council approved a plan to replace what was left of Henry McKenty's "swamp route" to Lake Como. Sadly, the colorful speculator was not around to witness this event. McKenty hanged himself in 1869 after falling ill and facing financial ruin. His widow received $5,000 in bonds from Ramsey County as repayment for his expenditures on the old Como Road.

By the mid-1880s, small neighborhood parks and squares within the St. Paul city limits numbered about twenty-five. Though Como Park remained a blend of meadows and forests, the surrounding areas were changing fast. The Northern Pacific Railway's maintenance center, the Como Shops, sprawled over land adjacent to the park property to the south, and due west the first Minnesota State Fair was held at its new fairgrounds at Como and Snelling Avenues, the former site of the Ramsey County Poor Farm.

Residential developments followed. Advertisements promised that "the value of the lots in Como Park Village will increase as rapidly as any suburban plat about the city, and in all probability will double within twelve months and quadruple as soon as Como Park is improved and opened as a pleasure grounds." Houses in the Warrendale development on the southwest shore of Lake Como started at $2,000, a base price that assured prospective buyers of "desirable" neighbors. The developers touted such amenities as a Presbyterian church, access to Lake Como, and something with a significance few at the time could have imagined: a large greenhouse.

Invited back for another round of consulting on St. Paul's parks in 1885, Horace Cleveland shrewdly focused his presentation on the surprising financial windfalls parks had brought New York, Boston, Philadelphia, Buffalo, and Chicago. Absent from his remarks were the old platitudes about nature and health. Instead he told how parks had dramatically increased the values of adjacent taxable properties, while curbing the flight of affluent residents from sections of the inner city with rising density and commercial activity. In a dramatic touch, Cleveland told the city council the story of the Roman emperor who foolishly declined the oracle's offer to predict Rome's future. Then he said,

The Sybil is immortal, and she continues to make the same offers to St. Paul that she made two thousand years ago to Rome. No flaming advertisements proclaim the value of her wares. Silently she displays them to us, and silently she departs when we decline the offer, and with her departure they are gone forever, for the name of Sybil is Opportunity.

By now Minneapolis had already organized a park board, elected Charles Loring president, and retained Cleveland as its landscape architect, a post he held for more than a decade. St. Paul's twin city was making quick progress in developing its Grand Rounds. In 1886, the state legislature authorized $25,000 in bonds for park improvements, and Horace

Designed with meandering walkways and lush flower gardens by landscape architect Horace W. S. Cleveland, Loring Park, originally called Central Park, was later named for its champion, prominent Minneapolis miller Charles M. Loring.

Como Park is owned by the City of St. Paul and managed by the Parks and Recreation Department, which oversees the 160 parks and open spaces that make up the St. Paul parks system. Spanning 384 acres, Como Park has 2.3 miles of paved walks and a 1.67-mile paved trail surrounding the lake. In addition to the Marjorie McNeely Conservatory, Como Park contains the Como Zoo, Como Town amusement park, the Cafesjian Carousel, Como Lake, an eighteen-hole golf course, a lakeside pavilion, a pool, picnic shelters, athletic fields, and numerous public art objects and gardens. Frederick Nussbaumer's "people's park" today welcomes more than 2.5 million visitors annually.

Cleveland moved from Chicago to Minneapolis. A rivalrous twinge likely factored into St. Paul's rush to action. The city council handed over control of its parks to the newly formed St. Paul Board of Park Commissioners, which immediately purchased land that would become Cherokee, Hiawatha, Indian Mounds, and Carpenter Parks and, over a period of five years, stitched together the parcels to create a park adjacent to Lake Phalen.

The long-delayed development of Como Park as a "landscape park . . . for the physical and moral sanitation" of the people of St. Paul finally began in 1887. The park commissioners hired Cleveland to prepare "such a design and plan . . . as he may think best suited to its topography."

The landscape architect did not take these orders lightly. His duty, he wrote, was to "serve as the high priest of Nature . . . to interpret her language, and to develop her suggestions . . . the symbols through which she addresses her worshippers." In his 1889 and 1890 designs, roads and pathways follow the undulations of the hills. The city workhouse, located since 1881 on forty acres at the southwest corner of the park site, was one of only three buildings. Formal plantings were limited to the drives along the park peripheries: Lake Drive, circling Lake Como, and Indian Point Drive, which cut between Lake Como and the much smaller Cozy Lake. McKenty Street (now West Jessamine Avenue) marked the park's southern border. Cleveland had the entire park seeded with short grasses, and a variety of shrubs and vines planted as ground cover. Trees would frame and enhance natural vistas.

Apart from the drives, a picnic ground was the only hardscaping completed by the time Cleveland's contract with St. Paul ended in 1890 and the project was handed over to a man who became even more attached to Como Park than Cleveland had been. Frederick Nussbaumer started at Como as a gardener. He quickly rose to become superintendent of all St. Paul's parks, a job he kept for the rest of his working career. Among his many contributions were the lavish floral displays that enticed more visitors than any other feature of the park in the ensuing decades and which were made possible by another of Nussbaumer's legacies, the system of greenhouses that sheltered both tender and tropical plants in the winter months. The centerpiece—the magnificent conservatory built in 1915—would become a national historic treasure.

The Como Zoo evolved informally through gifts at the turn of the nineteenth century. Its three deer lived briefly on Harriet Island before being moved into more spacious digs at Como Park in 1897. More animals followed, including foxes, elk, Zebu cattle, and the two buffalo given in 1915 by Lieutenant Governor Thomas Frankson. The animals roamed in outdoor pens, though the conservatory sheltered the less hardy of the lot on cold winter days. Charles Bassford designed the art deco Zoological Building in 1936. The opening of the Visitor Center in 2005 and Tropical Encounters in 2007 spotlighted the complementary relationship between the zoo and the conservatory at Como Park. Today the zoo collection features more than five hundred animals. It is one of four remaining free zoos in the country.

A serene Cozy Lake postcard scene from 1905.

CHAPTER THREE

the Tropics under Glass

BORN IN BADEN, Germany, and educated at the University of Freiberg in mechanical and civil engineering, botany, and floriculture, Frederick Nussbaumer honed his skills at the Royal Botanic Gardens at Kew, outside of London, and at a large nursery in France where he may have first met Horace Cleveland. Impressed by the young German's talent and enthusiasm, Cleveland advised his new friend to seek work in America. Nussbaumer did, settling in St. Paul, also probably at Cleveland's suggestion.

German gardeners were in high demand in many American cities. Frederick Law Olmsted's design for Central Park was executed largely by German immigrants. Whereas most American-born landscape architects were self-taught, men like Nussbaumer arrived with training and experience gained in the great gardens of Europe. Just as forestry and gardening jobs were well-populated by German laborers, a rich floricultural tradition also directed German immigrants to the expanding floristry business. Indeed, Nussbaumer's passion for flower design increasingly put his vision of the park at odds with that of Cleveland, for whom flowers, especially when arranged into man-made pictures and patterns, had no place in a naturalistic pleasure ground.

Conflict between those who liked floral displays (and recreational facilities) and those who preferred an entirely "natural" park was heating up in other cities as well, most notably in New York, where purists eventually succumbed to the formidable will of park superintendent Robert Moses. The debate over parks' proper role in public life continues today; the argument represents a timeless struggle for the soul of a landscape, a conflict loaded with philosophical, moral, and practical considerations.

In Como Park, the outcome of this intellectual debate was beginning to take shape. Substantial acreage in the northern and southern sectors of the park remained wooded, cut through by serpentine parkways. In line with the naturalistic aims of the English pleasure

A slice of the Como Park landscape, circa 1900.

ground, Nussbaumer planted more oak, maple, elm, birch, and lilac. He also experimented with non-native smokebush, sweet shrub, and white fringe, and even a few marginally hardy species like Camperdown elm, tulip tree, magnolia, and beech. A golden willow grown from cuttings was said to be a descendent of a tree on the site of Napoleon's grave on St. Helena. Flowers and shrubs edged many of the dirt and gravel walks that cut through broad lawns to allow for wide views. Willows followed the curve of the Willow Walk, while Glen Pathway ran through a landscape of shorter shrubs, conifers, and planted flowers. Silver maples and silver poplars lined the lakeshores. Tall oaks shaded the picnic grounds on Hamline Avenue.

The Willow Walk in Como Park, circa 1908.

In the Great Meadow, west of Como Lake, parkgoers flew kites and pitched quoits. Flowers and tall grasses encircled the lily pond. Young lovers courted on Hogsback Slope near Cozy Lake. Strolling, bicycling, picnicking, or just lounging on the grass were common pastimes. With George Eastman's invention of the hand-held box camera, Como Park became a mecca for amateur shutterbugs. Photographers were allowed to take pictures in the park as long as they didn't sell them to the public. The Haas Brothers photography studio's "official" images of Como Park appeared in the park board's 1895 annual report.

Park rules prohibited shooting, launching fireworks, and climbing trees. Picking flowers and eating fruit from the trees were also citable offenses. Any horses, cattle, sheep, goats, or geese found running loose were impounded. It was a far cry from the scattering of city parks in St. Paul, where cows had been set to graze.

Increasingly, modern urban parks like Como were acquiring pavilions, refectories, shelters, drinking fountains, and benches. No longer were hardy native plants considered the only media for artistic expression. Parks served as backdrops for works of art of a different type, in which plants were planted in elaborate designs that might be described as the opposite of natural beauty. These new plants were largely tropical, either annuals arranged in formal designs or spectacular specimens such as palm, fig, and banana trees that were as marvelous to winter-weary Minnesotans as the lions and tigers in the zoo that would soon become a popular park attraction.

Nussbaumer summarized his evolving views on park design in a 1902 essay called "An Ideal Public Park."

A successful or ideal park must provide facilities for recreation and, to a certain degree, objects of attractiveness in horticultural displays—especially so in the high northern latitudes where, on account of the long, bleak winters, the floral decorations in public parks excite special admiration during the short summer season.

Such displays, judiciously sited, would not "conflict with any landscape scenery, or with the general character of naturalness in the park except as to strict rurality."

Whereas Cleveland thought exotic floral displays would undermine his goal of "stimulating a poetic sensibility," Superintendent Nussbaumer was willing to forego

The novel Avenue of Palms was a favorite walkway for visitors to Como Park at the turn of the century.

the Arcadian ideal for the sake of stimulating less lofty but, to his mind, equally worthy and certainly more reliable and broadly accessible feelings. Moreover, he yearned to harness his training in floriculture to produce eye-popping wonders.

Flowers, he said, "pay for parks" because "[the] great mass of people enjoy flowers." An 1895 Como Park map shows a large area bordered by Midway, Autumn, and Lawn View Avenues earmarked for flower beds, which also graced the entrance to the park near the streetcar station. Elsewhere in the park, roses and peonies tempted walkers to stray off the paths for a whiff of their sweet scents. Even in the international heyday of English-style naturalistic landscaping, wherever grand floral displays were installed people flocked to see them.

Nussbaumer found inspiration for his designs in Europe. Flowering annuals arranged in artistic patterns that were either geometric or representational harked back to the formal gardens at Versailles. The ultimate example of nature under man's control, the designs made no reference to the origins of the individual plants in the design. Their colors, shape, and textures were all subordinated to their effect en masse, whether they depicted a flag, an emblem, or simply a pleasing abstract design.

The French technique called "mosaiculture" (the English called it "carpet bedding") actually originated in Nussbaumer's homeland, Germany. It was derived from Renaissance knot gardens and parterres. As explorers brought home more and more tropical plants from their travels, which greenhouses kept in cultivation, mosaiculture spread throughout Europe and across the Atlantic. Novel displays flourished in American municipal parks, on private grounds, and at fairs. Popular motifs included clocks, cornucopias, and commemorative civic designs incorporating town names.

In 1895, Nussbaumer's gardeners spelled out "Como" in flowers and surrounded the word with a

Elaborate and fragrant floral carpet bedding at Como Park, circa 1905.

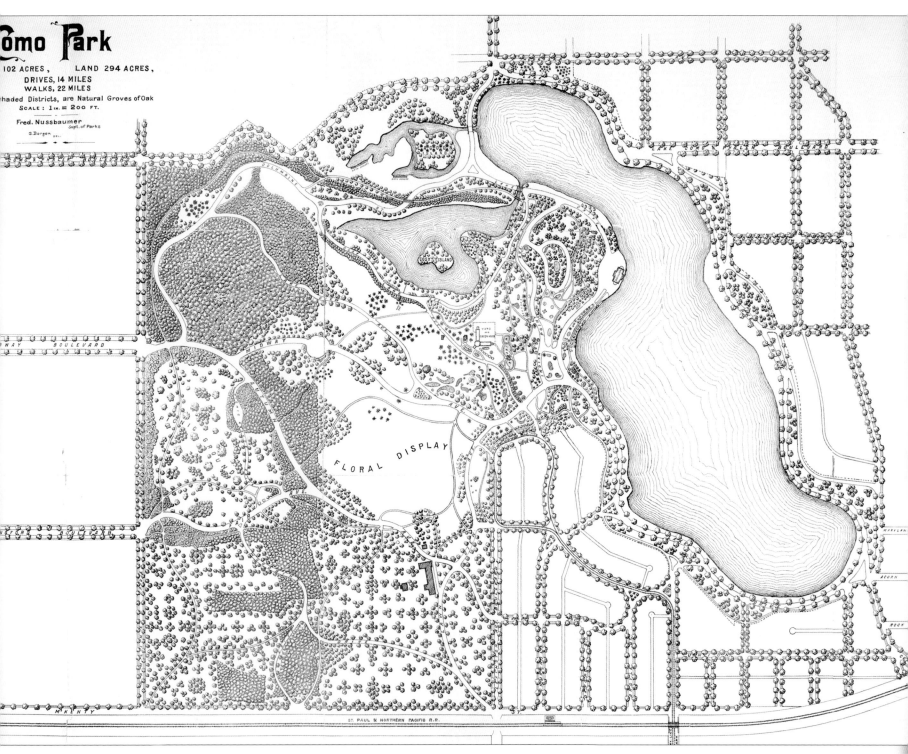

An 1895 map prepared by Frederick Nussbaumer shows the early landscaping of Como Park.

large horseshoe; an American flag and a star overlooked Cozy Lake. Leveling the ground and setting out the plants was an arduous process. Gardeners made use of colorful and sturdy ornamentals such as coleus, alternanthera, echeveria, centaurea, pyrethrum, sempervivum, and amaranthus. Complex designs could incorporate thousands of plants.

Whether tiny seedlings or spectacular palms, Como's tender plants required a winter enclosure. The annuals needed only a climate-controlled production greenhouse, but the big tropical specimens—the palms, figs, and bananas—begged for an exhibition space that would allow visitors to enjoy them in January as well as in July the way visitors who crowded the spectacular conservatory at Kew Gardens in England did. Recent advances in building materials and technology made the challenge irresistible.

The first attempts to grow exotic plants in climate-controlled spaces took place in Ancient Rome. Melons from Asia and Africa were produced in sunken enclosures glazed with semi-transparent sheets of mica. Tiberius Caesar had cucumbers, his favorite vegetable, grown on flat wooden carts that his slaves wheeled around from one sunny spot to another. The carts were parked over manure pits to keep them warm at night.

By the Renaissance period, citrus fruits were most esteemed. Maritime expeditions enriched the collections of European botanical gardens, where botanists strained to catalog the host of imported species. In the sixteenth century, wooden or stone structures heated by small flames protected the fashionable fruits from winter temperatures. Mobile wooden cold frames covered the trees.

The first fixed glass shelters appeared in the Netherlands and were built of stone, brick, or wood and were warmed by the midday sun shining through tall, paned windows. Wood- or peat-burning stoves added more heat. The utilitarianism of these early "orangeries" gave way to elegant Baroque structures heated by pipes that circulated warmth from a separate fire room built against a north wall or from stoves in the basement. As the collections expanded to include lemon, pomegranate, and myrtle trees, and a diversity of exotic plants, the term "greenhouse" entered popular parlance.

The Royal Botanic Gardens, Kew, established in southwest London during the 1770s, financed expeditions all over the world to collect exotic flora such as palms, cocoa, bananas, and rubber that could be studied for their commercial potential. By 1789, between 5,000 and 6,000 plant species grew at Kew. Specimens other than seeds, bulbs, dry rhizomes, and roots could not survive long ocean voyages, though, and in 1833 Dr. Nathaniel Ward, a London physician, introduced his "Wardian case," a forerunner to the terrarium that offered sunlight and humidity to plants set out on the deck of a ship.

Empress Josephine de Beauharnais, the wife of Napoleon Bonaparte, was France's grande dame of botany at the turn of the nineteenth century. Her garden at Malmaison boasted one of Europe's largest collections of rare flowers and exotic plants. She commissioned a greenhouse 164 feet long and 21 feet wide, with a slanting expanse of glazed wood frames and 12 coal-burning stoves.

Tropical plants found shelter under glass in homes throughout Europe, and not just in royal families. Middle-class families, too, attached conservatories to their homes much as we do garages today. Andrew Faneuil, a wealthy Boston merchant, built the first American greenhouse on record in 1737. He used it primarily to grow fruit. George Washington first bit into a pineapple in 1751, and later declared the chubby yellow fruit his favorite among the many tropicals grown in his red-brick hothouse at Mount Vernon. Its floor-to-ceiling windows shed ample light on the citrus trees, Sago palms, coffee,

The Palm House at Kew Gardens in London, designed by Decimus Burton and Richard Turner and built in 1844-1848, was 353 feet long, 100 feet wide, and 66 feet high and inspired conservatory builders worldwide.

and aloe that spent the hot Virginia summer outside in tubs. A below-ground heating system, like those once used in Roman bathhouses, brought Washington's greenhouse up to semi-tropical temperatures.

New materials and techniques made greenhouses larger, stronger, and more widely accessible. Wrought or cast iron enabled architects to engineer wider spans. Iron was frost- and water-resistant, and could be cast into decorative forms. By 1833, mass-produced sheet glass could measure up to six feet. Rolling the molten glass into even sheets reduced imperfections that concentrated the sun's rays and caused leaves to scorch. Iron and glass expanded the simple, square cube into seemingly limitless sizes and shapes. The repeal of Britain's Glass Tax in 1845 lowered prices and further expanded the realm of possibility.

Architects loved working with glass. Solid facades evaporated, revealing structural underpinnings well before the modernists of the early twentieth century made "form and function" a fundamental precept of their design philosophy. Glass buildings were light and transparent. Expansive vaults and domes appeared to float in the sky. Unimpeded by traditional building materials and customs, glasshouse builders experimented with new forms. Scottish botanist and greenhouse builder John Claudius Loudon tested conical, hemispheric, and cylindrical shapes that were beautiful and better able to admit light and dispel water.

Joseph Paxton, the Kew-trained gardens superintendent for the Duke of Devonshire, oversaw the construction of his design for a wood- and iron-framed conservatory at Chatsworth House. The Great Conservatory was a 277-foot long, 123-foot wide, and 67-foot high tent-shaped structure with one curved roof section sitting atop another. Cast iron pillars and columns supported a roof constructed of wood sashes holding 4-foot glass sheets in an innovative pleated pattern that allowed light to enter consistently throughout the day.

This was the largest glass house to date, but its innovative use of glass throughout is what set the Great Conservatory apart from its greenhouse predecessors. It took ten men to run the eight coal-powered boilers that heated the building through seven miles of iron pipe. Coal was transported to the site through an underground tunnel. One evening in 1842, the Great Conservatory was lit with twelve thousand lamps, and Queen Victoria rode in her carriage from one end to the other. She wrote in her diary that this was "the most stupendous and extraordinary creation imaginable." Lush vegetation of "gigantic proportions" filled the Great Conservatory at Chatsworth. Exotic birds flew freely and gold and silver fish flitted about in pools. The King of Saxony likened it to "a tropical scene with a glass sky."

The Jardin d'Hiver in Paris, built by Hector Horeau in 1848 on the Champs-Élysées, housed a café and pâtisserie, reading room, ballroom, waterfall, several fountains, a *jardin anglais* complete with a lawn, and hundreds of potted trees and plants. Steam heat kept the building comfortable through the winter. Mirrors seemed to double the space and the number of occupants. Best of all, unlike the Great Conservatory at Chatsworth, the Jardin d'Hiver was open to the public.

Between May and October of 1851, more than six million visitors from around the world crowded London's Hyde Park for the Great Exhibition of the Works of Industry of All Nations. The achievements of Britain and other invited nations were showcased in the Crystal

Joseph Paxton's Crystal Palace for the 1851 Great Exhibition of the Works of Industry of All Nations.

Palace—the name was coined by the satirical magazine *Punch*—designed by Joseph Paxton. Its iron framework supported more than one million square feet of glass, one third of England's annual glass production.

The term "conservatory" had long referred to greenhouses attached to homes. Now, however, the terms greenhouse, glass house, and conservatory were used interchangeably, though conservatory was always the name applied to glass structures lavish enough to host social events and provide romantic settings for strolling, playing cards, reading, chatting, or flirting.

Conservatories were important features of European parks when San Francisco's Golden Gate Park became America's first urban pleasure ground to boast a public conservatory. The building is now called the Conservatory of Flowers. Other American cities soon followed suit. The Lincoln Park Conservatory, built in stages between 1890 and 1895, had a palm house; a tropical house for flowering trees, vines, and bamboo; a show house for seasonal flower exhibits; and a sunken fern room. Industrialist Henry Phipps commissioned a conservatory in 1893 for Pittsburgh's newly developed Schenley Park. The Phipps Conservatory was the largest in the United States, with nine display rooms, a Palm Court, Serpentine Room, Fern Room, Orchid Room, and Victoria Room. Central Park finally got its flower conservatory in 1899. It was soon eclipsed by the New York Botanical Garden's fine new facility in the Bronx, constructed by Lord & Burnham in 1900.

Frederick Nussbaumer won approval for Como Park's first greenhouse for propagating bedding annuals in the fall of 1888. The basic wood and glass structure was soon doubled in size. Cold frames were added in early spring. By 1892, Como was supplying plants for all St. Paul parks. A new seventeen-by-forty-foot, lean-to style greenhouse, equipped with its own wood- and coal-burning stove, still didn't satisfy Nussbaumer's need for growing space. His detailed records show that Como gardeners raised 188,445 bedding and ornamental plants from seed annually. Of these, 151,890 flowering plants were installed at Como, more than half in the four carpet beds that had become the superintendent's pride and joy. He declared the 1897 displays "the handsomest of all." For several years, the Minneapolis park board was in the humiliating position of having to buy bedding plants from St. Paul, but in the late 1890s two permanent greenhouses finally went up in what is now Theodore Wirth Park.

When Lord & Burnham constructed James Lick's conservatory kit in Golden Gate Park in 1878, it became the first public conservatory in a United States park.

Constructed between 1890 and 1895, the conservatory in Chicago's Lincoln Park was a showplace and a working greenhouse that supplied plants for the park.

the Tropics under Glass

This 1920s postcard depicts the domed conservatory at the New York Botanical Garden in Bronx Park, New York.

In 1893, Lord & Burnham built the Phipps Conservatory in Schenley Park, Pittsburgh, Pennsylvania, for Henry Phipps, who gifted the building to the city.

St. Paul's Board of Park Commissioners in 1910. Frederick Nussbaumer is third from the right.

But it was California that became the nation's carpet-bedding capital, attracting visitors nationwide and stoking interest in the trend toward mosaiculture. Gardeners working in less temperate climates went all out in spring and summer to develop dramatic designs that would delight their audiences into the fall. The spirit of competition fueled rivalries between cities. When Friedrich Kanst of Chicago's South Park (German-born, like Nussbaumer) and Victor Siegel of Columbia Gardens in Butte, Montana, both introduced three-dimensional topiary in the form of a spherical planet Earth, Nussbaumer followed with his own version for Como Park.

Floral sculpture was essentially carpet bedding planted on wood and wire frames packed with moss and mud. Nussbaumer's first such experiment at Como Park, the 1894 green wonder called The Gates Ajar, featured a larger-than-life gated stairway made up of fifteen thousand hens and chickens (a perennial succulent groundcover) on a background of Joseph's Coat amaranthus (an annual grown for it multicolored leaves). The popular creation became a park fixture. A huge topiary elephant set on an island

"The Gates Ajar" stairway as it appeared in Como Park in 1898.

in Cozy Lake was Nussbaumer's 1895 encore performance. In 1896 a floral fort armed with green cannons and topped with a bald eagle honored the Grand Army of the Republic.

Nussbaumer knew his works bordered on kitsch. "Even such pleasant artifices of bedding plant composition as The Gates Ajar, The Globe, and The Elephant . . . are not, of course, designed for any purpose of artistic decoration, but simply to amuse the public," he declared in 1896. Nussbaumer's critics were not as sanguine. They complained that his designs were strewn haphazardly about the lawn and had nothing to do with their surroundings, that moreover the flowers looked "ugly and mean" when planted so tightly. Park board president Joseph Wheelock shot back that

> [p]arks are not merely pictures for the delectation of a few finical virtuosos. Their main purpose is to make the beauties of nature minister to the creation and enjoyment of the people at large—of the plain people whose pleasure grounds they are and to whose use and benefit they are dedicated.

Wheelock happened to be the editor of the *Pioneer Press* and a good friend of Nussbaumer's. A fervent advocate for park improvement, he appealed to civic pride through this broadside aimed at St. Paul's sister city:

> That the flowers in Como Park are a potent element in its remarkable popularity there is no question. Its visitors count about 1,300,000 during the park season. Large numbers of them come from the neighboring city of Minneapolis, whose flowerless parks . . . are poorly patronized by the public.

It was not long before Nussbaumer was asking for more greenhouses. In September of 1891, the park board had approved an addition to shelter "the big plants now in Como Park." Park board notes on the "big plants" mention no particulars, but the board members were most likely referring to the trees that lined the Avenue of Palms and Banana Walk during the summer.

Horace Cleveland deplored such theatricality. While Nussbaumer agreed with him in theory and wrote in praise of native plantings "in harmony with our climate, with the hue and tint of our clouds and skyline," in the same 1902 essay he admitted that such subtle, everyday charms might be lost on the average working person.

The inviting veranda on Lake Como attracted park visitors.

CHAPTER FOUR

Jewel in the Crown

THE EXPANSION of his greenhouses late in 1891 didn't stop Frederick Nussbaumer from pushing for more. Soon the park board was studying a set of plans furnished by architect Haas for a conservatory to be attached to a new residence for the superintendent and a steam heating plant for both. Nussbaumer's two-story house and attached conservatory were to be built west of Lake Como. In September, the park board accepted bids totaling $14,000, with the provision that work be completed by December 1, 1892.

By November 15, the greenhouse built by Lefebvre and Deslauriers was heated by steam piped in from the new plant built by engineer Allan Black. Contractor M. B. Farrel had Nussbaumer's house up by December 1 and heated by mid-month. The banana trees, palms, and a variety of other prized plants spent the winter of 1893 warm and cozy "under cover of glass."

An early photo shows banana plants growing in a wood-edged bed positioned in the center of the iron-framed hexagonal glasshouse, and potted plants sitting on raised wooden flats. The conservatory (later referred to as the Palm Dome) cost $8,310, twice the cost of Nussbaumer's house. Earlier park greenhouses had mainly been workspaces, but park visitors

Como Park's first conservatory was small but the plants were big.

By 1895, bananas grew in the first conservatory at Como Park.

were allowed to peek inside this one.

Visitors witnessed a horticultural first in Minnesota when Como Park gardeners coaxed the *Victoria regia* and its cousin, *Victoria randii*—water lilies native to the backwaters of the Amazon River—into bloom. This astounding event occurred in a man-made outdoor pond, often referred to as the "aquarium," that was heated in spring and fall "in order to bring out the splendor of the strange beauties." Nussbaumer had ordered the plants from William Tricker & Company. Tricker grew and hybridized water lilies in its Independence, Ohio, greenhouses. On arrival in Minnesota the plants were potted and submerged in the outdoor pond. The hybrid *Nymphea Devoniensis* joined *Victoria*, but the Amazonians attracted the most attention "on account of their monstrous leaves and flowers." Some of the wide platters, as the saucer-like leaves were called, measured four feet in diameter. The pineapple-scented flowers blushed from white to rose-purple over the course of their brief nocturnal bloom. Wooden footbridges crossed the pond to the "flower hill" on the opposite shore.

The aquarium was a crowd pleaser. Period photos show park visitors decked out in summer strolling costumes topped with straw hats standing shoulder to

The earliest accounts of Europeans witnessing the gigantic water platters of the Amazon go back to the turn of the nineteenth century. But it was not until 1837 that English horticulturist and botanist John Lindley described and named them *Victoria regia* (now *V. amazonica*) in honor of the reigning British monarch, Queen Victoria. Seeds and specimens sent to England sparked competition among horticulturists hoping to grow specimens whose leaves surpassed nine feet in diameter, with stalks three times that in length and flowers "larger than milk pans." In 1849, Joseph Paxton borrowed seedlings from the Royal Kew gardens and *V. amazonica* bloomed in a custom-built tank heated by coal-fired boilers to replicate the lily's warm, swampy habitat. *V. amazonica* flowered for the first time in the United States in the gardens of wealthy merchant Caleb Cope in Springbrook, Pennsylvania, in 1851. By the end of the century, the exotic lilies were a staple item at parks across the country. Como Park was no exception.

Placing small children on the platters of *Victoria regia* (*V. amazonica*) was a popular photo stunt in the 1890s.

shoulder on the bridge, gaping at the Victoria lilies. At least one hundred platters floated on the surface of the pond in the spring of 1896. Once again Como was on the leading edge of park floriculture. Though the lilies were difficult to cultivate, park gardeners grew them successfully through the 1910s.

After a brief hiatus, Como Park gardeners returned *Victoria amazonica* to the lily pond in the twenties, but by 1927 the elm trees surrounding the original lily basin shaded it completely. (In 2005, conservatory gardeners succeeded in growing from seed a less fussy hybrid, Victoria 'Longwood x Hybrid', a *V. cruziana* and *V. amazonica* cross developed at Longwood Gardens in Kennett Square, Pennsylvania. That summer they were on view in the heated outdoor pool that wraps around the new Visitor Center. The gardeners have since treated the lilies as annuals.)

Additions and improvements at Como Park came in a flurry over the next decade as the superintendent and his board responded to ever-higher public expectations. Park visitors clamored for a glimpse of the Japanese garden, new lakeside pavilion and promenade, picnic grounds, and water fountains. The board approved the purchase of a locomobile to carry Nussbaumer on his daily park rounds.

In 1909, the superintendent built a second lily pond (now called the Frog Pond) just south of the greenhouses. Park Board notes refer to this pond as the "Nelumbium Pond and Rockery." Nelumbium refers to the genus of Asian lotus, *Nelumbium speciosum*, grown in a basin constructed of 123 cords of limestone trucked in from Newport, Minnesota. The cost of the rock alone was over $1,500. Nussbaumer talked up the new pond as more than just another exotic attraction for the park. He told his board that

this improvement will prove to be of the highest importance to the visiting public, as it will aim to demonstrate the art of water gardening and rockwork on an extensive scale. [The pond will] furnish additional facilities for aquatic plant study to interested visitors.

The Como Park collection of palms and other tropicals was growing in size and number. Though two more small propagating houses had been constructed, Nussbaumer's palm house was seriously overcrowded. By 1913, the nine Como greenhouses supplying plants for the entire park system were badly in need of repair.

The first Japanese Garden at Como Park was located at the edge of Cozy Lake.

A seat on a bench was hard to come by on this day in Como Park, late 1910s.

Jewel in the Crown

Space constraints put plants at greater risk of disease and injury. When board president Chester R. Smith assembled his fellow commissioners on May 19 to address the problem, Nussbaumer seized the moment to propose that Como replace its motley collection of greenhouses with a single conservatory—a crystal palace like the domed marvels being constructed in city parks across America. For two decades he had been harboring a dream for such a sparkling shrine to floriculture at his park.

Some viewed the project as sheer folly, a waste of taxpayer money, a foolish extravagance. However, these critics were outmaneuvered by Nussbaumer supporters, who allocated fifty dollars for "a general set of plans for the contemplated erection of new greenhouses in Como Park." The determined superintendent had his proposal written by the end of July. It called for the destruction of all the existing greenhouses to make way for a new conservatory and ancillary production greenhouse "sufficient for the need of the department for the next fifty years."

The production greenhouses would provide bedding and ornamental plants for St. Paul's parks, while the conservatory would allow visitors to view the expanding botanical collection in an appropriately dramatic and exotic setting as close as possible to the plants' natural habitat. Nussbaumer further envisioned his crystal palace as an ideal site for educating elementary and high-school students in basic horticulture.

The project was included in the $280,000 bond issued that year for improvements to St. Paul's parks and parkways. Nussbaumer and the Toltz Engineering Company of St. Paul collaborated in the design process. King Construction Company of North Tonawanda, New York, submitted the winning bid to supply the structural elements for the conservatory. King Construction had recently built the largest greenhouse in the world in North Wales, Pennsylvania—three acres under glass and six miles of walks. King Construction's own manufacturing plant claimed the distinction of being the first factory ever made entirely of glass. More importantly, it sat at the intersection of two major railroads.

Spanning sixty thousand square feet, the conservatory (along with the three production greenhouses) arrived by train in St. Paul as a kit of prefabricated parts including the iron and steel framework, cedar ribbing with fitted curved glass, and redwood trimming. The bill matched the square footage almost exactly. King Construction's assembly team traveled to St. Paul to ensure that the kit was put together per their specifications.

Workhouse inmates graded the site and prepared the foundation at a cost to taxpayers of $15,000. They placed the cornerstone, containing a time capsule, in May, and erected and glazed the production greenhouses.

Then the steel and iron skeleton of the conservatory went up. There was snow on the ground by the time workers began fitting glass panes into the cedar ribbing. The following summer saw the completion of the glass dome and the addition of rot-resistant redwood trim, painted white. In style and spirit, the conservatory was directly descended from the tropical palm house at Kew Gardens. Classical balance organized the ground plan. A glazed skeletal structure conveyed lightness and elegance. Arched wings extended out on a north-south axis from a central dome composed of two superimposed units one hundred feet across at the widest point.

Hardy workmen glaze the conservatory through the winter of 1914-15.

Jewel in the Crown

Rising seventy-two feet from the ground to the tip of its ventilating cupola, the dome was the tallest part of the structure, suitable for displaying the tallest palms. Ionic pilasters decorated the spaces between its louvered windows. The motif carried over to the glassed-in vestibule at the main entrance. Three curved-eave production greenhouses extended from the east side of the central dome. A service building, boiler room, and coal shed were located in the rear. The building enclosed one-half acre under glass.

Four years after its completion, the jewel in the crown of St. Paul's park system sheltered 152 palms of

Workmen pose in front of the Como Park Conservatory, which they built in 1914-1915.

22 species. The collection included over 1,000 ferns, 90 orchids, 6 types of aquatic plants, and more than 130 species of greenhouse plants, including bananas. In the production greenhouses, 61,690 bedding plants were readied for transplanting throughout St. Paul's parks.

For park visitors, the conservatory was a botanical temple, offering a peaceful reprieve from the city, a year-round tropical paradise, and a safe haven for the plants that decorated the parks in the summer. For Frederick Nussbaumer, it was the realization of a twenty-five-year dream, and the greatest achievement of his life.

The Como Park Conservatory viewed from across the Frog Pond in 1916.

Horticulturist Dave Patsche installs floral arrangement in the pond in the Sunken Garden.

CHAPTER FIVE

Show time!

ON SUNDAY, November 17, 1915, the leaves had fallen and the chill of late autumn swirled through the air in St. Paul. Under the glass roof of the Como Park Conservatory the temperature averaged seventy degrees and the moist air smelled of warm earth and flowers. Superintendent Nussbaumer was on hand to guide park visitors on their maiden tour of the indoor exhibits. From late morning through late afternoon, more than three thousand people walked through St. Paul's shiny new crystal palace. An orchestra played a concert inside the building that afternoon.

Como Park's splendid new glass palace opened with much fanfare in November 1915.

Como Park's sightseeing bus in front of the conservatory in 1917.

One of the propagating greenhouses at Como Park in 1920.

Head gardener Adolf Kelper ran the greenhouses under Nussbaumer's direction. For the young gardener, it was trial by fire. In preparation for the event, sixty-seven chrysanthemum varieties were grown in the park greenhouses. Gardeners arranged the potted plants in the large hourglass bed and at the edge of the curving path in the aquarium. A sago palm and banana plant added height at the center of the design. Hundreds of mums placed on rough pine tables lined the straight parallel walkways of the conservatory and the growing houses at the rear. The *Pioneer Press* reporter covering the event wrote nothing about the conservatory's shimmering glass skin, the span of the steel framed dome, or the even the palm trees growing below it. What impressed the reporter, and the general public, were the chrysanthemums.

In the nineteenth century, flower shows enjoyed the same cachet among the cultural elite as any fine-art exhibit. The St. Paul Horticultural Society had been established in 1860. St. Anthony and Minneapolis garden clubs quickly followed. Annual exhibitions and smaller monthly competitions encouraged members to exchange knowledge and open their wallets. The first Minnesota Flower Show, held on July 4, 1863, netted a $600 profit. Prizes were usually

The moist interior of the Palm Dome during the summer of 1918.

modest, a ribbon or a small monetary sum.

By the turn of the century, mum shows were annual affairs at botanical gardens, conservatories, department stores, and societies. Chrysanthemums drew a "great throng of women and children, and a man sprinkled in here and there" to New York's Madison Square Garden for the "most successful flower show ever held in this country." Three years later, in 1890, a group of Chicagoans formed The Chrysanthemum Society of America. An exposition in that city featured mums from horticultural societies and flower clubs across the nation and Canada. It was billed as the "the greatest chrysanthemum show ever held in the world." The Missouri Botanical Garden hosted one hundred thousand for its annual chrysanthemum show in 1905.

The chrysanthemum was introduced into Japan

Show time!

from China around the eighth century. So smitten were the Japanese that only royal and noble families were permitted to cultivate the flower. Dutch traders brought mums to Europe in the late seventeenth century. After Imperial Japan was opened to trade in 1853, Japanese culture seduced Westerners with its exotic mystery, and chrysanthemums quickly became the flower of choice in Europe and America. In Victorian-era America, chrysanthemum bouquets were de rigueur for autumn brides. As large mums were a hothouse crop that had to be fussed over, they became associated with wealth. In 1895, white and pale pink chrysanthemums adorned St. Thomas Episcopalian Church in New York for Consuelo Vanderbilt's autumn marriage to the ninth Duke of Marlborough.

While Charles Sprague Sargent, the first director of the Arnold Arboretum at Harvard University, found the ubiquity of mums cloying, the 1892 issue of *Garden and Forest* predicted no end to their popularity, but rather "a return to a larger variety of types, and many of the old kinds will come into favor. Already we see this."

Indeed, the diversity of cultivars available made mums perfect for flower shows. Growers relished showing off new hybrids. Spider or quill mums could almost pass for sea anemones. Dedicated professional and recreational gardeners devoted months of intensive care to coaxing the so-called dinner-plate mums to gigantic size. A record-breaking crowd came to the American Museum of Natural History in 1912 to see "mammoth" mums from the gardens of local tycoons. "Bigger is better" was the mantra of the early twentieth century, and mums were no exception.

Frederick Nussbaumer stepped aside as superintendent of Como Park in 1921, leaving the park in splendid shape after thirty years. His successor, St. Paul landscape architect George L. Nason, declared his mission "to make the displays equal to the structure." He added flower shows at Christmastime, in mid-winter, and over Easter. He had a fourth growing house rebuilt and a fifth added in 1924. The conservatory was upgraded, with new marble walks installed in the North Garden (the aquarium) and Palm Dome. Nason turned the North Garden over to culinary herbs, medicinals, and other useful plants for what he called an "economic garden." Cacti took up residence in a corner of the Palm Dome.

Nason also added the Sunken Garden to the

Aquatic plants added texture, color, and drama to this 1920s flower show in the Sunken Garden.

conservatory's south wing in 1927. Upon entering the room, visitors stood on an overlook five steps above the garden. Below, flowers and a path lined a keyhole-shaped ribbon of water. Small lily pads floated on the water's surface and raised beds planted with flowers and small trees cloaked the outer walls of the room in lush greenery. Ferns hung from the iron trusses above. At the far end of the garden, a stone stairway carried the eye toward the pointed arch of the south vestibule, which framed a display of potted plants and flowers.

The Sunken Garden's first shows were all about floral abundance. Pots arranged around the pool sloped upward toward the outer walls, where flowering trees limited the view outside the glass walls, creating a dense,

Show time!

The giant exhibition chrysanthemums attracted this 1920s lovely lady.

contained effect. In 1927, precisely 5,374 plants of 263 varieties, 48 never before shown in Como Park, made up the first chrysanthemum show held in the new Sunken Garden. More than five thousand people inspected the flowers on opening day, and the conservatory stayed open until nine every evening for next three weeks to accommodate the crowds.

In 1929, cast cement benches were installed for guests to sit down and enjoy the spicy fragrance of hundreds of towering, single-stemmed exhibition mums during the chrysanthemum show. Huge hanging potted ferns and poinsettias in the 1930 holiday show were a far cry (and to some a welcome change) from the usual Christmas-in-Minnesota pageantry.

During the Great Depression, the conservatory fell into disrepair, and parts of it had to be closed. But the flower shows went on when people needed them most. During the spring of 1933, masses of potted tulips and lilies transformed the Sunken Garden into a floral dreamland, and beauty queens from four lands—Miss Paris,

Miss Paris; Miss Berlin; Miss Lima, Peru; and Miss Florence, Italy, pose for a photo op among the tulips in 1933.

Miss Berlin, Miss Florence, and Miss Lima—posed for pictures. The 1937 holiday show was highlighted by a giant faux pipe organ and rose window. A picturesque thatched English cottage was the centerpiece of the 1939 tulip show.

The development of a hardy chrysanthemum cultivar in the 1930s failed to dampen enthusiasm for the cherished, fall mum-show tradition. The 1936 show had all the staginess of the popular Hollywood Charlie Chan movies. The main attraction that year was a Chinese-style pagoda in the Sunken Garden. The pagoda had bamboo shutters, gilded ornamentation, and two ceramic dragon statues standing guard. Inside its illuminated entrance, single-stemmed exhibition mums grew upward like a regiment of lollipops. The giant mums spilled out and around the pagoda into the rest of the room, which was filled with flowerpots. A woman wearing Chinoiserie-style silk pajamas strolled about the room.

In 1936, the conservatory received a Gothic cathedral façade to mark that year's St. Paul Winter Carnival, as well as a faux stained-glass picture depicting

Show time!

King Boreas extending the "Fun Key of the Carnival." Four years later, the coronation of Boreas Rex VIII was staged on the conservatory steps, with two thrones set against a purple and gold backdrop. Colored lights trained on the steam rising from the dome gave the fairy-tale scene its finishing touch.

Elaborate commercial flower shows raised the bar for the Como Conservatory. (Macy's in California started the trend, and in Minneapolis Dayton's department store put on its first show soon after.) The conservatory had to keep up. The Easter show in the Sunken Garden lured thousands to see the spring lilies, tulips, hydrangeas, hyacinths, and narcissis, while the Easter Bunny entertained children and posed for photos.

A faux façade for the conservatory during the 1940 St. Paul Winter Carnival.

A scaled-down version of Anne Hathaway's cottage decorated the 1939 Spring Flower Show.

At Christmastime, plastic snowmen stood amid the cyclamen and poinsettias at the holiday flower show.

The fiftieth annual Como Park Conservatory Chrysanthemum Show, in 1968, was an especially elaborate affair. Japan supplied the theme. The Men's Garden Club partnered with the St. Paul-Nagasaki Sister City Committee. The Sunken Garden was transformed into a Japanese-style wonderland. A wooden bridge crossed over the water to a temple bedecked with rice-paper lanterns imprinted with kanji characters. Garden club members and others garbed in Japanese *happi*, or "happy coats," strolled about. Three St. Paul residents born in Nagasaki visited the show in silk kimonos and posed for photographs. In one of the greenhouses, a few inches of raked

Como Park Conservatory gardener Joe Chenoweth seems pleased with the grapefruit harvest grown in the North Garden in 1951.

Children from the Wheelock School in St. Paul were awed by the Palm Dome in 1941.

white sand and large black rocks introduced the public to a Zen garden complete with a red torii gate, bonsai trees, Japanese flower arrangements, and what the *Pioneer Press* described as "a place with pillows on the floor around some kind of Japanese broiler." The show was such a hit that it was repeated in 1973.

Como superintendent Robert Schwietz enthused: "Not only do we get the usual wild profusion of chrysanthemums including oriental varieties with crazily gesticulating petals, but we get educated." Schwietz ran Como Park for twenty-six years, beginning in 1952. He raised attendance figures and dollars by courting garden clubs and working closely with the Minnesota Horticultural Society. His goal was to transform the Como Park Conservatory from "just a showplace into a modern botanical garden."

Meanwhile, the flower shows proved irresistible to anyone with a camera. Mothers coaxed their families to the gardens for Easter-bonnet photos. Donaldson's department store shot clothing ads at the conservatory, and visiting beauty queens always posed for photographs there, surrounded by flowers. "Snow Queen Out of Her Realm" headlined a series of photos showing Eva Wicker, the recently crowned St. Paul Winter Carnival

A young miss leans in for a whiff of the chrysanthemums.

Parents show one of the big mums to their baby girl during the 1957 Chrysanthemum Show in the Sunken Garden.

Show time!

Kimonas on stage, 1968 Chrysanthemum Show.

Queen of the Snows, cavorting among the palms, fashioning a sarong out of a banana leaf, and playfully pitching a grapefruit at the photographer.

Just as they had in the late 1800s, the floral splendor and dense interior jungles of conservatories appealed to mid-twentieth-century romantics. Schwietz once arranged an after-hours marriage proposal for twenty-five-year-old Jack Harper of White Bear Lake. In a scene worthy of Hollywood, Harper drove Jean Svendsen to the conservatory three hours after closing time. A park policeman let the couple in, turned on the lights, and then busied himself for twenty minutes while Harper popped the question to Svendsen in the Sunken Garden. Happily, she accepted.

Schwietz once noticed that an elderly couple had been sitting a long time under a blossoming loquat tree. He engaged them in conversation. The couple explained that they'd courted under a loquat tree in their native Syria. Today was their wedding anniversary, they told him. The superintendent was so touched that he admitted feeling tempted to break a cardinal rule of the conservatory and pick one of the ripe fruits for them to take home.

For Como's fifty-third annual chrysanthemum show, themed "Hours in Flowers Soon Fade Away," a giant floral clock designed in the carpet-bedding style greeted guests in the Sunken Garden. The clock measured nearly ten feet across, its massive hands stuck at a quarter past eight. But the "hours in flowers" never did fade away in the eternal summer of the Como Park Conservatory. No sooner was the clock removed than the Christmas poinsettia show was up and running.

Even today, the conservatory still feels removed from the passage of time and from the realities of the everyday world. It's a place where winter coats come off, a place that always smells of warm, damp earth and flowers.

Ethereal and dreamlike during a winter snow in 2009, the Marjorie McNeely Conservatory promises pleasures within.

Show time!

The Century Plant breeched the glass ceiling before erupting in hundreds of yellow blooms during the summer of 1963.

Early one day in the spring of 1963, the unexpected appearance of an asparagus-like spear signaled to the conservatory gardeners that a rare event was in the making. The spear was the flower bud rising from the center of *Agave americana*, commonly known as the "Century Plant." Native to tropical parts of the Americas, especially Mexico, *Agave americana*, is a large succulent with a spreading rosette of gray-green leaves that have spiny edges and sharp tips.

The plant is monocarpic, meaning that it flowers only once and then dies. Although the number of years it takes each plant to flower varies between five and sixty, *Agave americana*, was dubbed the Century Plant because it stores nourishment in its fleshy leaves for years before producing buds. Actually, the size rather than age of the plant determines when flowering will occur, and the conservatory's Century Plant had been growing its rosette for many years.

In 1963, no one could remember exactly when the plant had arrived at Como Park. The 1930s was the best guess.

On June 23, 1962, hailstones the size of golf balls smashed through the glass roof of the Como Park Conservatory. Chaos ensued as glass shards pelted visitors who had come inside the building to escape the storm. The bombardment forced the gardens to close for the first time. The plant collection also suffered extensive damage. Gardeners worked diligently to care for the wounded plants and trees, donning special helmets to protect themselves from the glass pieces that continued to fall from above. An emergency appropriation of $75,000 was made to replace the broken panes with fiberglass. Public festivities marked the project's completion in October and the November mum exhibit played to enthusiastic crowds. The show was held over for an extra few days.

A devastating hailstorm in 1962 left the Como Conservatory in this sorry condition.

The Big Storm of '62

Visitors entered the Como Conservatory through its historic front entrance in the late 1950s.
The dome was painted with a shading compound to reduce heat and excessive sunlight.
Plywood panels on the adjoining North and Sunken Gardens protected their glass from ice falls off the dome.

CHAPTER SIX

another Rescue

IN 1974, the Como Park Conservatory was placed on the National Register of Historic Places. This proud moment in the history of the conservatory was well-timed, because the building needed help. Years of exposure to the elements had caught up with previous renovations, among them new glass and supporting ribs in the Palm Dome and flanking display houses installed under the auspices of the Works Progress Administration; aluminum to replace rotting wood; and a new boiler. Now the building's metal supports were rusting, the foundation was crumbling, heating pipes needed replacing, and the whole structure badly needed a fresh coat of paint. Fiberglass panels installed after a hailstorm had shattered the glass panes in 1962; now the fiberglass was clouded, cutting off sunlight to the plants within.

The Como Conservatory Restoration Society, formed in 1971 to finance major repairs, found it harder to raise money as the decade wore on and an economic recession lingered. When Marvin (Les) Day became conservatory manager in 1978, the city's parks budget had been slashed and several Como employees had been laid off. Desperate to keep the growing houses in operation, workers wrapped them in polyethylene and replaced older furnace burners. Still, annual operating costs held steady at $250,000. By 1980, Day was considering shutting down some of the growing houses and switching to hardier plants.

Como's wasn't the only conservatory facing hard times; Victorian-style palaces had become an endangered species. Milwaukee lost its 1898 Victorian conservatory in 1955. Three modern, beehive-shaped domes replaced it. The 1902 conservatory at the New York Botanical Garden was slated for demolition when, like Como, its listing as a historic landmark in 1973 helped launch a rescue effort. The $5-million, two-year project, underwritten by philanthropist Enid A. Haupt in 1978, was followed by another in just fifteen years, this one costing $21 million.

San Francisco's Conservatory of Flowers had experienced a never-ending roller-coaster ride. A boiler

Pictured here about 1910, San Francisco's Victorian-era, one-of-a-kind Conservatory of Flowers was assembled from a pre-fab kit in 1878.

explosion set the main dome on fire in 1883, burning it down. The dome was restored, survived the earthquake of 1906, burned again in 1918, and was rebuilt a second time. Due to structural instability, the building was closed between 1933 and 1946. Much of the dome was repaired in 1968 and again in 1978. In 1995, a fierce windstorm destroyed the dome yet again. Faced with multimillion-dollar repairs, San Francisco's parks board considered razing the redwood relic. Redemption came when the World Monuments Fund placed the boarded-up building on its list of most endangered world monuments, prompting the National Trust's Save America's Treasures and First Lady Hillary Rodham Clinton to adopt the building in 1998. The $25-million renovation of America's first public conservatory was completed in 2003.

One hundred thirty years young, San Francisco's conservatory in its renewed glory.

another Rescue

In such hard times, Les Day had his work cut out for him. Raised on a South Dakota farm, Day held a degree from South Dakota State University along with horticultural experience gained while working for Johnson Wholesale Florist in Roseville, Minnesota. He was the right man at the right time for the venerable but down-at-the-heels Como Park Conservatory. Practical and optimistic, the new superintendent encouraged the public to take ownership of the building. "We all need something to care for," he said. Looking beyond the peeling paint and algae-encrusted windows, Day saw a major horticultural resource that needed to hone its mission. At his urging, the St. Paul Men's Garden Club and the Bromeliad Society of Minnesota made the building their regular meeting spaces. The Orchid Society of Minnesota, the Minnesota Bonsai Society, and the Minnesota chapter of Ikebana International were invited to put on shows. The Minnesota Horticultural Society formed a Como Park Conservatory Preservation Society to mobilize support and raise funds. Day proudly noted in 1980 that visitors to the conservatory had increased over the previous two years. In 1982, the first groups of Como Park Conservatory volunteers were trained to help with the plants, tours, and fundraising.

A conservatory gardener for twenty-three years recalled that as soon as Day arrived, the gardeners started getting better equipment and more supplies. The care and quality of the plant specimens improved. Day expanded the plant collection, added more growing houses, and even built an onsite soil-sterilization facility. He encouraged the gardeners to brainstorm on displays. Asked what he had learned from his years in commercial horticulture, Day offered a comparison. In his former work he was judged by "the almighty dollar," whereas at Como "it's how much your spectator is pleased." He clearly preferred the latter form of compensation.

Day's can-do attitude energized his staff, his volunteers, and the community, and impelled them to take seriously the challenge of halting the conservatory's physical decline. Some city officials thought bulldozing the old building was the only option, but the majority was persuaded to rescue the grand old building. Day asked volunteers to develop a master plan for renovating and rehabilitating the building. City of St. Paul landscape architect Bill Pesek agreed to serve as project manager.

Taking as a given that the original design would be retained in accordance with its historic landmark status, committee members pored over the 1915 blueprints

The Como Conservatory at its down-the-heels worst needed rescuing.

and studied similar renovations in other cities. The plan also had to address community education and recreation. In 1984, after consulting with architects from Ralph Burke and Associates and the Division of Parks and Recreation, the Como Conservatory Planning Advisory Committee published the Como Conservatory Master Plan. The detailed, 280-page roadmap won initial funding from the Metropolitan Council. The first phase of a $13.5-million project started a year later.

Over the next seven years, the architecture firm Winsor/Faricy restored the Como Park Conservatory from the bottom up. Heating, ventilation, water, and electrical systems were also slated for repair or replacement, funded by a $500,000 city council appropriation.

Construction uncovered a mess of unforeseen problems. The steel supports and concrete support walls showed serious corrosion. Some of the inner metal walls were so rusty a pencil could be poked straight through them. Wooden ribs were loose and glass panes had cracked in the Palm Dome. The Metropolitan Council

another Rescue

transferred $800,000 intended for Como Park and the Como Zoo to the conservatory restoration program in 1986, the first expenditure put toward emergency repairs to the Palm Dome's collapsing walls.

Even with the building under siege, the conservatory's annual flower shows still drew big crowds that year. For the spring show, gardeners arranged more than twelve thousand tulips, daffodils, and hyacinths in the Sunken Garden. The wildly popular exhibit evoked French Impressionist painter Claude Monet's garden at Giverny. Day took cuttings in June for the three thousand mums representing more than one hundred varieties that would make up "Showtime '86," an extravagant display of flowers, including scenery, props, and costumes from the Chimera Theater of St. Paul.

The opening of the fall mum show kicked off a major fundraising campaign for the conservatory. Mayors George Latimer of St. Paul and Donald Fraser of Minneapolis launched the Como Conservatory Restoration Fund, with contributions from corporations and individuals amounting to $1.7 million in its first two years.

In 1989, work on the Sunken Garden put the mum show on hold until the following fall. Conservatory improvements included blue-tinted glass, fresh paint, two elevators that made the split-level room accessible for visitors with special needs, new iron railings and flagstone walks, and updated climatic and electrical systems. Construction had also begun on a state-of-the-art production greenhouse complex consisting of eight planting bays.

In the Palm Dome, North Garden, and Fern Room, plants that had never even produced buds began flowering, and avocado trees bore fruit. Manager Day was thrilled. "It's giving the place more of that crystal palace feeling," he said. A special $3.53-million bond issue carried the work through to the Como Park

Ripe, plump oranges in the North Garden.

Palms and cycads have thrived in the historic Palm Dome since its inception. With an ascending roof height of approximately sixty-four feet at its apex, the glass rotunda is well suited for this iconic tropical family of plants.

Conservatory's seventy-fifth anniversary year, and in March 1991, the Sunken Garden reopened with a spectacular show of five thousand fragrant tulips.

Two years (and $2 million) later the framework and glass of the conservatory's most recognizable feature, the Palm Dome, had been entirely rebuilt. Day felt his work was done. He had come to the conservatory when it was falling apart and left it a structurally sound building with state-of-the-art growing facilities. In September 1993, he turned over the job of conservatory manager to Roberta Sladky, who held the position for thirteen years.

A graduate of the University of Wisconsin-Madison with a masters degree from the University of Delaware in public horticulture administration, Roberta Sladky was no stranger to Como Park. She'd left a position as the Minnesota Zoo's horticulture curator to head up the conservatory staff. A volunteer gardener in 1985, she now supervised seven gardeners

another Rescue

The Common Fig (*Ficus carica*) in the North Garden.

and upwards of two hundred volunteers. Arriving at the completion of one renovation, she jumped into another almost immediately.

As conservatory manager, Sladky hoped to revive the study of nature as a leisure activity for families whose free time was increasingly taken up with passive indoor pursuits. Plant identification and interpretation were her first initiatives. New signs and labels not only identified plants by name but offered detailed information on their cultural origins.

In the fall of 1995, Sladky introduced the Environmental Education Department. A grant from General Mills brought in a local artist-in-residence, who assisted students in creating art inspired by the plant collections. The Merrill Corporation funded a weekly noon-hour concert program called "Music Under Glass." Gardening classes resumed, some taught by Sladky herself, others by volunteers, including many retired seniors. Volunteers also gave guided tours. An after-school program offered hands-on learning to children from St. Paul's McDonough Recreation Center, who created art using plants and animals from different cultures.

It was time for Como Park Conservatory to throw itself an eightieth birthday party. Children made

A single volunteer managed to raise more than ten thousand dollars by selling Como-themed handicrafts to conservatory visitors. Terry Stanke was a trained horticulturalist who for a time had run a greenhouse with her husband. She'd visited the conservatory frequently as a child and became a volunteer in 1962, making flower arrangements for the mum show. In 1988, she came up with the idea of recycling fallen flower petals into boutique novelties. As the staff cleaned up at the end of each day, Stanke scooped up the discarded petals and took them home, dried the petals on her dining room table, added scented oils, and packaged the sweet confetti in old pimiento jars. Stanke knew that Dayton's sold a similar product for eight dollars. "So why couldn't we make our own potpourri—and raise money for a good cause?" Each jar of Como Conservatory Potpourri was finished with a satin ribbon, a plastic pumpkin, or holly sprig. The conservatory's gift shop sold each jar of potpourri for one or two dollars.

Another Stanke idea was to propagate scented geraniums (Pelargonium is the Latin genus name) to benefit the conservatory. Three centuries of hybridization had created a vast array of scents, ranging from baked goods to Martha Washington's famous rotting fish flowers. In the 1930s, blind persons came just to smell and touch the scented geraniums, Stanke recalls. "They lost a lot of the flowers during restoration," she says. "I decided it was my job to bring them back." With help from the conservatory gardeners and her own professional connections, Stanke was able to search out enough cuttings to revive the conservatory's stock of scented geraniums. She sold twenty varieties in the conservatory gift store—lemon, apricot, and chocolate mint, among them.

Terry Stanke's volunteerism spanned more than 40 years at the conservatory, from 1962 until 2004.

another Rescue

flower masks, painted clay pots, planted ferns to take home, and made floating water lilies. In the Sunken Garden, Sunday visitors enjoyed concerts ranging from jazz to barbershop quartets. Thursdays brought art history lectures exploring the representation of flowers in art. The Historic Jubilee included music, theater, trolley rides, antique autos, vintage fashions, and a parade.

Fittingly, the finale of the birthday celebrations marked the November 7 opening of the chrysanthemum show and Ikebana display. The conservatory was packed. Kids tramped up and down the steps, squealing, smelling, and touching. Water pools turned into wishing wells were carpeted with pennies, dimes, quarters, and even pesos. Older visitors reminisced: many recalled riding to the park on the street car. Others told of falling in love at Como Park. In fact, countless Minnesotans have tied the knot at the conservatory, and tens of thousands have chosen the conservatory as the backdrop for formal photography shoots from weddings and anniversaries to proms and holiday portraiture.

Others reminisced about more unusual experiences. One man told Sladky he'd stood in line as a college student in 1963 to see the century plant bloom. In April of that year this plant, an *Agave Americana*, had been Como's star attraction as it grew "through the roof." A section of the North Garden roof was removed to allow its flower stalk to grow until it was ready to bloom. Floodlights had illuminated the plant at night so the curious could monitor its progress, and a local television station offered "a C-note" to the person who correctly guessed the number of blooms (three thousand) it produced, until, on July 1, the spike snapped in a gust of wind.

When the newly restored Grand Veranda, Frog Pond, and plaza opened to the public, romance was the theme of the festivities. The stone lanterns in the Como Ordway Memorial Japanese Garden were lighted for the first time the following week. Honoring excellence in historical design, plant collection, and educational programs, the American Society of Horticultural Science gave the conservatory its Horticultural Landmark Award in 1999. Thomas Jefferson's gardens at Monticello and the U.S. Botanical Gardens were the only prior recipients.

During the mid-1990's, an effort was undertaken to evaluate the most effective structure for managing the conservatory and its neighbor, Como Zoo. Mayor Norm Coleman partnered with Paul Verret, then president of

Carrie Dittmer and Mark Carey exchange vows during the Holiday Flower Show in the Sunken Garden. The conservatory hosts more than 350 weddings and special events each year.

another Rescue

The Saint Paul Foundation, to convene a citizen group to research options and make recommendations. Tight city budgets and a desire to move forward with capital improvements necessitated a new administrative structure for the facilities that sat side-by-side, but had different hours of operation and separate staff for maintenance, education, and facilities management. Further complicating the issue were the existence of four separate non-profit organizations that supported different elements of the conservatory or zoo.

The group recommended the conservatory and zoo be managed under one administrative structure and move towards shared programs and staff. It further recommended that all existing non-profit organizations be merged into one organization that would represent both the conservatory and zoo. Thus, in April 1999, Como Friends (originally the Como Zoo and Conservatory Society) was incorporated as the sole non-profit fundraising organization for the conservatory and zoo. Later that year, Como Park Zoo and Conservatory merged into a single administrative unit.

Charged with seeking funds from the private sector for capital improvements, expanding programs, growing an endowment, and best practices operations, Como Friends immediately began raising money through a variety of initiatives: major giving program; membership program; animal sponsorship program; gifts, grants and sponsorships from corporations; three special events (Zoo Boo, Sunset Affair and Bouquets Wine Tasting); and the gift shop.

The administrative realignment paved the way for the undertaking of a long-dreamed-of visitor center with improved amenities and space for classrooms. In March 2000, Mayor Coleman announced plans for the first major initiative of the combined Como Park Zoo and Conservatory and its partner Como Friends. A $32.41-million capital campaign was launched with five objectives:

1. Construct a Visitor Center to provide public amenities such as a year-round café, expanded and accessible restrooms, a centrally located gift shop, and a main entrance. The Visitor Center would also have classrooms and an auditorium for education programs and after-hours rentals.

2. Construct a new wing of the Conservatory to replace a growing house converted into the Fern Room, which was no longer structurally sound. This new wing would house a new Fern Room as well as a state-of-the-art space for the orchid collection, a children's gallery for experiential learning, a bonsai gallery, and traveling exhibits.

3. Build Tropical Encounters, a new habitat that would blend animals and plants in one space, combining the missions of the zoo and conservatory, and allow visitors to experience a rainforest ecosystem.

4. Construct an Animal Support Building to provide cutting-edge facilities for animal quarantine, diet preparation, and animal training as well as improved facilities for the birds, reptiles, and amphibians previously housed in a basement space.

5. Renovate the 1930s-era Zoological Building to provide badly needed office space for zoo and conservatory staff.

Four years later, the conservatory and zoo had successfully secured $24.5 million from the Minnesota State Legislature through bonding appropriations in 1998, 2000, and 2003. Como Friends was successful in the first major fundraising effort it took to the community, securing $7.91 million in private gifts.

The years following witnessed one opening celebration after another, as Como Park Zoo and Conservatory and Como Friends introduced the new facilities to the community: the Animal Support Building in June 2002, the Visitor Center in February 2005, the new wing of the conservatory in May 2005, and Tropical Encounters in November 2006.

The new Visitor Center was designed by the architectural firm of Hammel, Green, and Abrahamson.

HAPPY FACES

RUDOLPH SCHIFFMAN, the patent-medicine tycoon-turned-Como Park patron, was not the first wealthy American to fall under the spell of Japanese garden design, which he discovered in 1904 at the Louisiana Purchase Exposition in St. Louis. A Japanese garden at the 1876 Philadelphia Centennial Exhibition had so impressed art collector and philanthropist Isabella Stewart Gardner that she commissioned one for her estate. In San Francisco's Golden Gate Park, a five-acre Tea Garden boasted ponds and a waterfall, imported trees, plants, bronzes, koi, rare Japanese birds, a Shinto shrine, lanterns, and a wooden Buddha.

Dr. Schiffman was so awed by the Imperial Japanese formal garden designed by Yukio Itchikawa, a landscape architect for the Mikado, that he purchased a "rare and large collection" of trees, shrubs, and granite lanterns from the St. Louis exhibit, and hired Itchikawa to assemble these treasures on the south-facing slope of Cozy Lake. Itchikawa's design featured cherry trees, stepping stones, and a bridge. A waterfall made of tufa rock spilled into two pools before dropping into the lake. Six-foot granite lanterns were placed at various scenic points adjacent to the water.

Though the garden was popular immediately, it could not withstand Minnesota's harsh climate, despite the presence of the small greenhouse annex that had been erected to shelter the plants. The garden racked up more than $1,000 in maintenance bills in its first year and then vanished from park records, its memory lingering on until a second—and this time permanent—Japanese garden was installed several decades later.

Shortly before retiring as conservatory supervisor in 1978, Robert Schwietz began to lay plans for a new Japanese Garden to highlight the twentieth anniversary of the St. Paul-Nagasaki sister city affiliation. Nagasaki's mayor had given Schwietz plans for the garden in 1974, drawn up by Masami Matsuda, Nagasaki's landscape architect, and Mikio Tanguchi, the city's chief of parks. Schwietz presented the plans to St. Paul's parks officials, who took the idea to the philanthropy-minded Ordway family. Like Henry McKnight, the Ordways had benefited from 3M Company's remarkable success and saw the new garden as a way to express their gratitude to people of St. Paul.

Charlotte Ordway, in particular, loved plants and flowers. She had been a founder of the St. Paul Garden Club and a passionate horticulturalist. Her four children donated the garden in her memory, with the expressed wish that the Como Ordway Memorial Japanese Garden be used to further enhance the sister city friendship program between St. Paul and Nagasaki. As park supervisor, Schwietz had fostered cultural ties through the chrysanthemum shows. He delightedly embarked on the project.

St. Paul parks officials boasted that the new garden would be one of only eight authentic Japanese gardens in the United States. Ray Zierden, a Sacramento landscaper whose previous work at the park included the Hamm Memorial Waterfall beside the lakeside pavilion, was hired to construct the garden to Matsuda's specifications on one acre just north of the conservatory.

Matsuda designed in the Sansui mountain-and-water style. He envisioned the garden as a peaceful retreat displaying harmonious relationships among water, stones, and trees. The basic elements were to represent a miniature Japan—an island country of mountains and

Floating lanterns at dusk in the Japanese Garden for the Lantern Lighting Festival.

Kabuki dancers at the 2002 Lantern Lighting Festival.

rivers surrounded by the sea. An island accessed by a lovely wooden bridge was shaped like a turtle. Partially submerged rocks evoked fish.

Only plants that were proven hardy in Minnesota would be used. That was the hard lesson of Como's first Japanese garden. Como's second attempt was not without its own difficulties. Japanese tradition is unbending on certain design points. When Matsuda visited St. Paul in 1984, he found, to his dismay, that

> [T]he basic precepts of placement of rocks were ignored. The rocks are all living things. So, how to bring the 100-percent best out of the rocks is the challenge through the technique of placement. We have to speak to the rocks, know the spirit of the rocks. A traditional Japanese garden treats rocks and trees and shrubs as living things and respects them as a human being.

Matsuda noticed at once that many of the two hundred rocks in the Como Ordway Memorial Japanese Garden were showing their "sad faces." He meant the rocks were displaying their less physically attractive side. Some had to be dug up and rolled over to show their "happy faces," while others were dug up and moved to eliminate straight lines and triangular formations. In contrast to the formal, geometric gardens of Versailles, Japanese gardens celebrate nature's mystery. They convey the message that nature holds the key to enlightenment, and that man must live in harmony with it rather than attempt to tame it through logical reasoning (symbolized by geometric forms) or brute force.

Matsuda and Bill Pesek, St. Paul's city parks landscape architect, commenced a trans-Pacific dialogue to remedy the errors and to repair damage caused by five years of harsh Minnesota winters. Matsuda returned to St. Paul in 1989 to begin the $250,000 renovation and again the following year to supervise its execution, living in a trailer parked adjacent to the conservatory. He hand-picked rocks

Lighted paper lanterns recall Japan's annual Obon holiday, when families pay respects to their ancestors.

from local quarries, made sure all the boulders were placed happy-face-up, reordered stones set in too-straight lines, saw that vegetation was planted in irregular triangles and that trees were pruned with their branches spiraling upwards.

A teahouse built to reflect a sixteenth-century Japanese farmhouse was added to the design, its construction paid for by the St. Paul Branch of Zonta, a women's organization. The deteriorated concrete lining of the pond was also restored. Now Matsuda was ready to place six stone lanterns about the garden. Four had come from the 1904 Japanese exhibition in St. Louis, relics of the long-ago Japanese garden on Cozy Lake.

For the Minnesotan construction workers, building the garden was a cultural learning experience, if frustrating at times. "If it's close enough, it's close enough," one worker griped. Just as it takes years of training to design a Japanese garden, appreciation and understanding must also be acquired over time. A garden that bends the art's rigorous rules ends up a travesty from which one cannot learn. St. Paul's new garden is not only beautiful to the trained and untrained observer alike—it is also authentic.

"Matsuda San put a lot of himself into the project," Pesek said when the Como Ordway Memorial Japanese Garden officially reopened in May 1992. "There was a lot of effort, going seven days a week, to see that everything came out right this time."

happy faces

Water gardens at the Visitor Center entrance.

CHAPTER SEVEN

the Gift

CONSTRUCTION OF THE NEW WING of the Como Park Conservatory was made possible by one woman's passion for flowers and gardening. Marjorie McNeely passed away in 1998 at the age of eighty-two. The wife of St. Paul business leader and entrepreneur Donald McNeely and the mother of five children, she found time to engage in numerous cultural activities in the Twin Cities. The first woman to sit on the Guthrie Theater's board of directors, Mrs. McNeely was a member of the Friends of the Minneapolis Institute of Arts and an active member of the Minnesota Historical Society. In addition, for many years she volunteered to read books aloud for the Radio Talking Book Network, Minnesota State Services for the Blind. Bespeaking her love of flowers, she was also a former president of the St. Paul Garden Club, for which she led flower-arranging classes.

"She always had a garden and it was always

Marjorie McNeely

Renaming the Marjorie McNeely Conservatory.

flourishing with fragrant flowers that she enjoyed immensely," her husband Donald remembered. Her son Greg recalled her fondness for lilacs, perennials, and cut flowers. Artful arrangements decorated every room of their Manitou Island home in White Bear Lake, where Marjorie's garden was awash with color from spring to fall. Greg McNeely relished the family's frequent winter-month trips to the conservatory when he was a boy. "Our family used to come here in the middle of winter to get a breath of fresh air."

The McNeelys moved to Pebble Beach, California, in 1965 but retained their Manitou Island home and connections with St. Paul. When Donald McNeely and his family were contemplating a memorial tribute for Marjorie, the city where she had raised her family and to which she had given so much seemed like the ideal place. In December 2002, the Donald McNeely family made a $7-million gift for the Como Park Conservatory. The gift was the joyful culmination of two years of negotiations with Greg McNeely representing his family, Roberta Sladky connecting the mission of the conservatory to the interests of the family, and behind-the-scenes involvement from leadership at Como Friends, Space Center Inc., Saint Paul Parks and Recreation, and Mayor Randy Kelly's office.

This donation is the single largest private gift for a public structure ever received by the City of St. Paul and Como Friends. Renamed the Marjorie McNeely Conservatory at Como Park, the venerable structure would immediately receive $3 million from the Donald McNeely family to help finance urgently needed improvements, including the construction of the new wing. This was the first addition of a wing to the conservatory since it was constructed during 1914 and 1915. The remaining $4 million would be given over a twenty-year period and placed in an endowment to finally secure the Marjorie McNeely Conservatory for posterity.

The Visitor Center opened in February 2005. This 65,000-square-foot building was the culmination of

The new Visitor Center opens into Como Park Zoo.

Completed in 2005, the contemporary-style Visitor Center (left) leads into the historic 1915 Victorian-style conservatory.

the Gift

twenty years of planning, fundraising, and dedication to improving visitor amenities and classroom space for visitors at Como Park Zoo and Conservatory. Designed by Hammel, Green, and Abrahamson, the Visitor Center draws the zoo and conservatory together both structurally and visually, serving as the primary entrance to both facilities. With the iron and glass of the conservatory and the golden-toned, Kasota-stone facade of the zoo building carried over into the new structure, it was intended to enhance rather than compete with the original 1915 crystal palace. The sharp angularity of the futuristic addition, its facets twisting and bending to maximize winter sunlight and minimize summer sunlight, plays off the dome's smooth curves. The white steel framework and ribbon of shallow ponds also strike notes of continuity.

The Visitor Center houses public amenities as well as classrooms, meeting rooms, and event rooms. Additional classrooms on the upper level open onto a rooftop garden. An auditorium seats up to three hundred people. Guests can grab a meal or enjoy a cup of coffee at the Zobota Café. Garden Safari Gifts offers souvenirs and a variety of plant- and animal-themed gifts, while lockers, an ATM, restrooms, and a service desk accommodate visitor needs.

Two "tree-house" classrooms cantilever into the canopy of Tropical Encounters, the dramatic, jungle-like habitat for flora and fauna off the Visitor Center lobby. Connecting visitors with the global ecosystem is a primary objective of Tropical Encounters, where they embark on a guided adventure through a Central and South American rainforest. BIOS, an exhibition design company from Bainbridge Island, Washington, created the habitat, where cocoa, banana, peach palm, balsa, and an ever-growing number of vines, trees, and herbaceous tropical plants thrive in the dense forest environment. A fallen "canopy giant" tree brings bromeliads and orchids down to eye level. Such parasitic plants, called epiphytes, normally grow at least one hundred feet above ground.

Children press their faces against built-in aquariums and stare, wide-eyed, at the freshwater stingrays, river turtles, fish, and a thirteen-foot green anaconda named Jenna. Terrariums allow the public to view boas, poison dart frogs, and tarantulas. A glass-front wall showcase displays the world of leaf-cutter ants. While some insects in the habitat are simply to be viewed, others have serious environmental work to do as natural pest controls, a practice that began at the conservatory in the mid-1990s. Saffron

Como Park Zoo and Conservatory attracts two million visitors annually.

finches fly through the forest canopy, where Chloe the sloth sleeps throughout the daylight hours. Along the trail interpretive signs, resembling field notes, from a team of Minnesota research scientists explain the sustainable relationships among plants, animals, and humans in the rainforests.

Near the end of their stroll, visitors come to a sustainable rainforest farm, whose type is disappearing as a result of encroaching cattle operations and other development. The holistic ecology of Tropical Encounters demonstrates how plants and animals work together in nature. It also represents an unprecedented collaboration between Como's zoo and conservatory.

A glass-enclosed walkway running east from the main entrance of the Visitor Center leads visitors into the new wing of the Marjorie McNeely Conservatory, where the HGA-designed Bonsai Gallery, Orchid House, and Fern Room opened in the spring of 2005. The Leonard Wilkening Children's Gallery, plant production rooms, and restrooms are also part of the addition.

Staff horticulturists curate one of the top bonsai collections in the United States. The Marjorie McNeely Conservatory Bonsai Collection has received trees styled by Mas Imazumi and John Naka, nationally revered

Trident Maple trained five years as a bonsai.

bonsai masters. Most of these stunning miniature treasures had been cared for behind the scenes. A seasonal Bonsai Gallery made it possible for these ancient trees to be enjoyed by the public.

Adjacent to the Bonsai Gallery are state-of-the-art production greenhouses. The Orchid House, viewable through its glass exterior, is filled with plants destined for the Palm Dome, North Garden, and Tropical Encounters. Begun in the 1980s, the conservatory's orchid collection consists of neotropical or New World species and naturally occurring hybrid crosses representing some 900 different species.

Conservatory orchids compete each year in the Winter Carnival Orchid Show in association with the Orchid Society of Minnnesota and at the Minnesota State Fair. Among countless ribbons and trophies the orchids have brought home since 1998 are more than ten prestigious awards from the American Orchid Society. One certificate of cultural merit allowed the conservatory to give the name "Como's Magee" to all orchids cloned from the *Rossioglossiym insleayi* while the most recent certificate of horticultral merit was awarded in 2005 to *Anguloa hohenlohii* 'Marjorie McNeely Conservatory'.

Under a twenty-five foot ceiling in the new Fern

Room, staghorn ferns cling to a forty-foot wall of lava rock, and rare tree ferns native to New Zealand grow in warm mist so thick it sometimes clouds the vision of visitors circumnavigating the room. Other unusual varieties include a bird's-nest fern, heart's tongue fern, and silver-dollar maidenhair fern. Tall bamboo-like giant horsetail (*Equisetum myriochaetum*) as well as tiny spike-moss (*Selaginella moellendorffii*) thrive in niches in lava rock boulders. A Mee Fogg system supplies ultra fine mist to maintain optimal humidity levels and to modify

This state-of-the-art orchid production facility houses a significant portion of New World species and hybrids.

the Gift

An ever-expanding collection thrives in the Orchid House.

temperatures during the summer months. Photovoltaic cells in the glass roof, a project component underwritten by Xcel Energy, provide dappled light that replicates a natural forest environment while generating supplemental electricity for the campus's power grid.

Implementing sustainable design in the new addition not only reduces the ecological impact of the building, but also raises public awareness of solutions to environmental problems. A network of steel ribs and gutters on the leaf dome channels rainwater into pools below that send cool, moist air into the building during the summer months. Renewable cork and fiberboard cover the walls. The exposed, sealed concrete floors contain a high percentage of fly ash, an inorganic inflammable material produced from coal combustion in electric generating plants and usually disposed of in landfills. Counters are a local product made of fabric, recycled paper, and resin.

Horticulture Supervisor Karen Kleber Diggs manages the tropical plant collections within the support facilities and the conservatory. An orchid specialist, Karen oversees staff horticulturists in the curatorial work of the permanent botanical accessions, the biological pest control programs, and plant education programs.

The Tropical North Garden is based on the botanical economic garden model to showcase plants with culinary or medicinal utility. Here visitors discover the idiosyncrasies of the plant that chewing gum comes from, the sources of spices like allspice and cinnamon, bananas, coffee, dwarf oranges, and figs. If a visitor wants to know what accounts for those leaf clusters sprouting off the thick short branches of the fig (*Ficus carica*) tree, volunteer interpreters have the answer: The tree is pruned annually to keep it to a manageable size; a technique called pollarding ensures the tree never bears fruit, which some may consider a benefit as the figs smell terrible. Thanks to constant pruning, the tree is a masterpiece of form and texture. Its lovely, twisting, truncated branches with their dense leaf clusters are well-worth the price of a few lost flowers and foul-smelling figs. Here and there around the North Garden, illustrated signs offer tidbits of such information, often whimsical, about particular plant species.

Four plant families at Como—cycads, orchids, ferns, and bromeliads—represent a focused effort to create world-class collections. The cycads thrive in the dappled light of the Palm Dome's understory. Their thick, knotty trunks sprout umbrella-like canopies of slender dark green leaves that grow straight up but eventually bend under their own weight in a graceful downward curve. The tender new growth is like caviar to white flies. Time to bring on the ladybugs! Fighting insects with insects is one of many practices that reflect Como's continuing effort to stay abreast of environmentally sound horticulture. Other examples include growing more native plants in its outdoor gardens (and eliminating all remaining invasive species like buckthorn), composting all discarded plant materials, keeping spotless the conservatory's expanses of glass, and pruning large specimens to maximize available sunlight.

Showcasing one of the conservatory's most valued and long-held plant collections, the new Fern Room, in addition to fern plants, features the most diverse assortment of tree ferns found under glass in North America.

In the production greenhouses, Horticulture Supervisor Paul Knuth manages the scheduling and cultivation of plant material for both the Sunken Garden Flower Shows and outdoor gardens on the Como Campus and select areas of the City of St. Paul parks. Under Paul's direction, the production staff grows thousands of plants from seed or cuttings. Succulents, marigolds, ageratum, petunias, impatiens, begonias, and others are among the bedding plants that fill the outdoor gardens. Azaleas, chrysanthemums, poinsettias, Easter lilies, and spring bulbs are mainstays of the staff-designed flower shows.

The Sunken Garden attracts individuals looking for a lovely, quiet place to hang out with a favorite book, as well as families who relish sharing all this beauty with loved ones. A more romantic atmosphere is difficult to imagine. It's been the scene of many a marriage proposal and is a popular venue for anniversaries. As Frederick Nussbaumer knew so well, flowers are magnets for people.

With the completion of the expansive renovation and a secure financial endowment in place, an enviable legacy, Roberta Sladky left the Marjorie McNeely Conservatory in 2006 to take a public garden director position at Olbrich Gardens in Madison, Wisconsin. Following a national search for candidates, Tina Dombrowski became the new horticultural manager at the Marjorie McNeely Conservatory. Recruited from a public garden in Dallas, Texas, Tina had more than two decades experience in public horticulture. A graduate of Cornell University with a major in Floriculture and Ornamental Horticulture, she moved to St. Paul to assume leadership of a well-positioned, professionally staffed, and generously supported botanical conservatory and gardens.

A second aquatic garden opened at Como in 2008. More than fifty-five species of submerged, marginal, and floating plants now grace the exterior of the Visitor Center, where they show off a range of textures, colors, and sizes of aquatic and bog plants in an artistic design certain to inspire the adventurous home water gardener. Species featured included hibiscus, rain lily, creeping Jenny, and sweet flag, as well as papyrus, which originates from the Nile and was used for ancient paper-making.

The Victoria water platters' new summer home is the large pool on the south end of the Visitor Center, just outside the Tropical Encounters exhibit. The water platters share the pool with other tropical aquatic plants, including rice and water lettuce. Horticultural staff secured seeds from the water platters at Longwood Gardens, the late Pierre du Pont's estate-turned-public

Tina Dombrowski, Horticulture Manager, Marjorie McNeely Conservatory.

garden in Pennsylvania.

Such exchanges are part of the romance of being associated with a world-renowned botanical institution like the Marjorie McNeely Conservatory. Many of its rare specimens come to Como with a personal connection: a former colleague sent cuttings; a local gardener brought some seeds back from a trip to Africa. Also contributing to Como's vitality over the years is the sister-city program with Nagasaki, which has created lasting relationships with people and plants.

In April 2008, a Titan arum or corpse flower (*Amorphophallus titanum*) that grew from seed given to the Marjorie McNeely Conservatory by a University of Wisconsin-Madison professor who happens to be the world's foremost authority on the species, burst into its notoriously stinky bloom. A web cam was hung over the plant for a live performance online. The story of Como's rare achievement event made the nightly news.

In 2008, efforts began to move forward with the twenty-year Como Master Plan for expansion and improvements in the outdoor gardens, using a modern framework to interpret the historic gardens of the early 1900s. The first conservatory project to be developed under the master plan will be the construction of the Japanese Garden Experience. This multi-million-dollar project will include a new wing of the Marjorie McNeely Conservatory with a year-round Pavilion for displaying bonsai trees, a Garden and Terrace for seasonal outside bonsai display, and a realigned entrance through a Pine Grove Walk to the Como Ordway Memorial Japanese Garden.

Since Frederick Nussbaumer's dream of a crystal palace in Como Park came to fruition in 1915, the building has traveled a long road from dazzling to derelict and back again. With public and private support, and through the dedication of many, St. Paul's conservatory has expanded and reemerged in its golden years as an environmentally conscious, state-of-the-art

Horticulturist Joan Murphy working on Bonsai.

Horticulture Supervisor Karen Kleber Diggs.

BEHIND *the* SCENES

Horticulture Supervisor Paul Knuth.

Horticulturist Jill Heim trimming palms.

behind the scenes

horticultural and educational facility. The conservatory's skeleton of steel, iron, and redwood, and its skin of glass made it an engineering marvel in the Victorian era, part science and part art, a world far removed from the one just outside its doors and from the bustling city just a few streetcar stops away.

The streetcars are long gone, but beneath the soaring Palm Dome, primitive cycads and stately palms still tower over bromeliads and other exotics that look nothing like the plants grown in the average Minnesota garden (in contrast to the lovely but gentler aesthetic of the park landscape surrounding it). Annual flower shows entertain and delight, fountains spout and trickle, and the scent of rosemary and jasmine wafts through the air. All who come find comfort in a warm place on a chilly day. Today, Como Park's jewel shines brighter than ever.

Bromeliad hybrid (*Guzmania 'Orangeade'*) in the Palm Dome.

Bromeliad foliage takes many different shapes.

Santa Cruz water lilies (*Victoria cruziana*) in the Visitor Center pool at sunset.

PICTURES *from an* OPENING

On the afternoon of May 17, 2005, conservatory staff and volunteers celebrated the opening of the new wing of the Marjorie McNeely Conservatory with special guests and donors. Speeches of welcome and gratitude were extended by St. Paul Mayor Randy Kelly and members of the McNeely family.

pictures from an opening

Illustration Credits

THE BILTMORE COMPANY
Asheville, North Carolina
p. 32, Frederick Law Olmstead, oil on canvas, 91 3/8" x 60 3/4", Artist: John Singer Sargent, 1895.

TIFFANY JOHNSON BIDLER
Columbia Heights, Minnesota
p. 27, Harriet Whitney Frishmuth's *Play Days*, 2007.

CHUCK JOHNSTON
Afton, Minnesota
p. 4, Sunken Garden.

MARJORIE MCNEELY CONSERVATORY
St. Paul, Minnesota
p. 2, water lily. Photographer: Jackie Sticha; p. 6, lily. Photographer: Jackie Sticha; p. 10, Visitor Center. Photographer: Tina Dombrowski; p. 24 (left), Alonso Hauser's *Reclining Nude #5*; p. 60, Como Conservatory. St. Paul Parks and Recreation photograph archive at MMC; p. 63 (bottom), Como Park, ca. 1910. St. Paul Parks and Recreation photograph archive at MMC; p. 68, spring flower show. Photographer: Tina Dombrowski; p. 69, Como Conservatory, ca. 1933. St. Paul Parks and Recreation photograph archive at MMC; p. 73, 1920s flower show. St. Paul Parks and Recreation photograph archive at MMC; p. 74, Como Conservatory, ca. 1920; p. 75, Como Conservatory, ca. 1933; p. 77, 1939 Spring Flower Show; p. 78 (left), Como Park Conservatory gardener Joe Chenoweth, 1951; p. 81, Winter 2009. Photographer: Tina Dombrowski; p. 82, Century Plant, 1963. St. Paul Parks and Recreation photograph archive at MMC; p. 83, hailstorm, 1962; p. 84, Como Conservatory. Photographer: Jack Bailey, ca. 1958-62; p. 89, Como Conservatory; p. 90, oranges in the North Garden. Photographer: Don Olson; p. 91, Palm Dome. Photographer: Don Olson; p. 93, Terry Stanke, 2009. Photographer: Tina Dombrowski; p. 95, wedding. Photographer: Anderson's Designer Portraits; p. 97, Visitor Center. Photographer: Xavier Ortiz; p. 99, lantern lighting ceremony. Photographer: Don Olson; p. 100, Kabuki dancers. Photographer: Green Tea Productions; p. 101, lantern lighting ceremony. Photographer: Don Olson; p. 102, Visitor Center entrance. Photographer: Xavier Ortiz; p. 105 (bottom), Visitor Center. Photographer: Matt Wehner; p. 107, visitors. Photographer: Michelle Furrer; p. 114, Tina Dombrowski. Photographer: Lorie Jonas; p. 115, horticulture staff. Photographer: Tina Dombrowski; p. 116, Bromeliads; p. 117, Visitor Center pool. Photographer: Tina Dombrowski; p. 118-121, opening celebration, 2005. Photographer: Stan Waldhauser Photo/Design.

DONALD MCNEELY FAMILY
p. 103, Marjorie McNeely.

MINNESOTA HISTORICAL SOCIETY
St. Paul, Minnesota
p. 14, Henry McKenty's house. Photographer: Whitney's Gallery, Stereograph ca. 1863; p. 15, Aldrich's Hotel, Lake Como. Photographer: Charles Alfred Zimmerman (1844-1909), stereograph, ca. 1870; p. 20, water grotto, Como Park, 1895; p. 21, Mannheimer's Memorial, Como Park. Photographer: Louis D. Sweet. Postcard ca. 1905; p. 23, Lucille Kube at Como Park fountain, ca. 1930; p. 24 (right), astrolabe; p. 26, Paul Manship's *Indian Boy and His Dog*, 1967, *St. Paul Pioneer Press*; p. 37, Loring Park. Postcard ca. 1908; p. 39, Buffalo, Como Park Zoo, 1928; p. 40, Cozy Lake Bridge, Photographer: Sweet. Postcard, 1905; p. 42, Como Park ca. 1900; p. 43, The Willow Walk, c. 1908, V. O. Hammon Publishing Company, Minneapolis and Chicago; p. 45, Avenue of Palms. Postcard ca. 1905; p. 46, Floral Par-Terre. Photographer: Truman Ward Ingersoll. Postcard ca. 1905; p. 47, map of Como Park, prepared by Frederick Nussbaumer, 1895; p. 55, members of the Como Park Board. Photographer: Haas & Wright, 1910; p. 58, Lake Como. Postcard ca. 1904; p. 61, lily pond; p. 63 (top), Japanese Garden. Photographer: Sweet, 1905; p. 65, Como Park Conservatory, 1914-1916; p. 66, Como Park Conservatory, 1914-1916; p. 67, Como Park Conservatory. Photographer: William J. Hoseth, ca. 1916; p. 70 (top), sightseeing bus. Photographer: Lee Brothers, 1917-1920; p. 70 (bottom), Como Park greenhouse. Photographer: Charles P. Gibson, ca. 1920; p. 71, Como Conservatory, August 22, 1918; p. 76, Como Conservatory, ca. 1940; p. 78 (right), students from Wheelock School. Photographer: *St. Paul Dispatch & Pioneer Press*, 1941; p. 79 (top), 1957 Chrysanthemum Show. Photographer: *St. Paul Pioneer Press*; p. 79 (bottom), 1957 Chrysanthemum Show. Photographer: *St. Paul Pioneer Press*; p. 80, 1968 Chrysanthemum Show. Photographer: *St. Paul Pioneer Press*.

NEW YORK PUBLIC LIBRARY
New York, New York
p. 18, Bryant Park crystal palace, lithographed by Nagel and Weingärtner for publisher Theodore Sedgwick, John H. Levine Collection; p. 31, Pierre Martel's bird's-eye view lithographic image of New York's Central Park, 1864. Lithographed by J. C. Geissler and printed by Henry Eno for the Central Park Publishing Company, I.N. Phelps Stokes Collection of American Historical Prints.

LEIGH ROETHKE
Minneapolis, Minnesota
p. 12, The Schiffman Fountain; p. 17, Great Conservatory at Chatsworth, postcard; p. 22, Schiller Monument; p. 28, Como Lake; p. 38, Como Park; p. 49, The Palm House at Kew Gardens, postcard; p. 51, Joseph Paxton's Crystal Palace, 1851. *Dickinson's Comprehensive Pictures of the Great Exhibition of 1851: From the Originals Painted for H.R.H Price Albert*, 2nd ed. London: Dickinson Brothers, 1854; p. 53 (top), James Lick's conservatory; p. 53 (bottom), Chicago's Lincoln Park conservatory; p. 54 (top), conservatory at the New York Botanical Garden in Bronx Park, New York, postcard ca. 1920s; p. 54 (bottom), Phipps Conservatory, postcard 1893; p. 56, *The Gates Ajar*, postcard 1898; p. 86, Conservatory of Flowers; p. 87, Conservatory of Flowers; p. 92, North Garden.

RAMSEY COUNTY HISTORICAL SOCIETY
St. Paul, Minnesota
p. 34, Horace Cleveland.

ROBERTA SLADKY
Madison, Wisconsin
p. 25, *Crest of the Wave*; p. 104, renaming conservatory; 105 (top), Visitor Center; p. 108, Trident Maple bonsai; p. 109 (top), Orchid House; p. 110, Orchid Production; p. 112, Fern Room.

UNIVERSITY OF MINNESOTA LIBRARIES
Minneapolis, Minnesota
p. 19, King Construction cover, Stock Plan Books Collection, Architectural Archives.

Index

Bold/italic numbers indicate an image.

Adler, Otto, 14
Aldrich Hotel, 14, **15**
Aldrich, W. B., 34
American Museum of Natural History, 72
American Orchid Society, 108
American Park and Outdoor Art Association, 21
American Society of Horticultural Science, 94
Anderson, William R., 25
Animal Support Building, 97
Aphrodite, 23, **23**
Arnold Arboretum at Harvard University, 72
Astrolabe dial, **24**
Avenue of Palms, **45**, 57

Banana Walk, 57
Banning, William A., 35
Bassford, Charles, 39
BIOS, 106
Black, Allan, 59
Bonaparte, Napoleon, 43, 49
Bonsai Gallery, 107-08
Bromeliad Society of Minnesota, 88
Bryant Park, New York, 18
Burbank, James C., 33
Burton, Decimus, 17, 49

Cafesjian Carousel, 38
Calhoun, Samuel, 33
Caponi, Anthony, 25
Carey, Mark, **95**
Carpenter Park, St. Paul, 38
carpet bedding, **46**, 46, 55
Catholic Daughters of America, 24
Central Park, New York, 16, 30, **31**, 41, 52
Century Plant, **82**, 82, 94
Chenoweth, Joe, **78**
Cherokee Park, St. Paul, 38
Chimera Theater of St. Paul, 90
Chrysanthemum Show, 26, **74**, 74, 78, **79**, **80**, 80, 94, 99
Chrysanthemum Society of America, 71
Cleveland, Horace William Schaler, 16, 21, 23, 33-34, **34**, 36-39, 41, 44, 57
Clinton, Hillary Rodham, 87
Cochran Park, 26-27
Cochran, Thomas, 26
Coleman, St. Paul Mayor Norm, 94, 96
Como Conservatory Restoration Society, 85
Como Conservatory Restoration Fund, 90
Como Friends (formerly Como Park Zoo and Conservatory Society), 96-97, 104
Como Ordway Memorial Japanese Garden, 94, **99**, 99-101, 108, 114
Como Park, 11, 19, **21**, 21, **22**, 22, 23, 27, 35-36, 38-39, **42**, **43**, 44, **45**, 45, **46**, **47**, 52, 55, **56**, 57, **60**, 60-62, **63**, 64, **69**, **70**, 72, 74, 79, 82-83, 91, 94, 98, 104, 116
Como Park Conservatory, 25, 26, **66**, **67**, 69, 76, 79-80, **83**, 83, **84**, 85, 88, **89**, 89, 92, 103-04
Como Park Conservatory Preservation Society, 88
Como Park Zoo, 38, **39**, 39, 90, 94, 96-97, 106
Como Road, 15, 36
Como Town Amusement Park, 38
Courthouse Square, St. Paul, 16
Cozy Lake, 39, **40**, 44, 48, 57, **63**, 98, 101
Crosby, Frederick, 23

da Bologna, Giovanni, 23
and *Mercury*, 23
Day, Marvin (Les), 85-86, 88, 90-91
and Johnson Wholesale Florist, 88
Dayton's Department Store, 76, 93
de Beauharnais, Empress Josephine, 49
and Malmaison, 49
Diggs, Karen Kleber, 111, **115**
Dittmer, Carrie, **95**
Dombrowski, Tina, 113, **114**
Downing, Andrew Jackson, **29**, 29, 30

Eastman, George, 44
Elephant, The, topiary, 57
Emerald Necklace, Boston, 33
Environmental Education Department, 92
Excedra, 23

Fairmount Park, Philadelphia, 33
Faneuil, Andrew, 49
Fern Room, 90, 96, 107, **112**
Farrel, M. B., 59
Fillmore, U.S. President Millard, 30
Fjelde, Jacob, 22
and *Minerva* at Minneapolis Public Library, 22
and *Hiawatha and Minnehaha* at Minnehaha Park, 22
Folwell, William Watts, 33
Frankson, Thomas, 39
Fraser, Donald, 90
Frishmuth, Harriet Whitney, 25
and *Crest of the Wave*, **25**, 25
and *Play Days*, 25-26, **27**
and Rodin, 25
Frog Pond, 62, **67**, 94

Gardner, Isabella Stewart, 98
and Philadelphia Centennial Exhibition, 98
Garden Safari Gifts, 106
Gates Ajar, The, 55, **56**, 57
General Mills, 92
Gilbert, Cass, 20-21
Glen Pathway, 43
Globe, The, 57
Golden Gate Park, San Francisco, 33, 52, **53**, 98
and Conservatory of Flowers, 52, **86**, 86, **87**
Grand Veranda, 94
Grant, President Ulysses S., 33
Great Conservatory or Stove, **17**, 50-51
and Duke of Devonshire, 17, 50
and Chatsworth House, 50
Great Depression, 74
Great Exhibition of the Works of Industry of All Nations, **51**, 51
Great Meadow, 44
Great Northern Railway, 15
Grotto Fountain, **20**, 20

Harriet Island, 39
Haas Brothers, 44
Haas, architect, 59
Hammel, Green, and Abrahamson, 97, 106
Haupt, Enid A., 86
Hauser, Alonso, *Reclining Nude #5*, **24**
Heim, Jill, **115**
Hiawatha Park, St. Paul, 38
Hogsback Slope, 44
Horeau, Hector, 51
and Jardin d'Hiver, 51
Holiday Flower Show, **95**
Horticultural Landmark Award, 94

Hyde Park, 51

Ibsen, Henrik, 22
Ikebana International, 88
Indian Mounds Park, St. Paul, 38
Indian Point Drive, 39
Irvine, John R., 16
Irvine Park, 16
Itchikawa, Yukio, 98
and Mikado, 98

Japanese Garden, 62, **63**, 101
see also Como Ordway Memorial Japanese Garden
Japanese Garden Experience, 114
Jefferson, Thomas, Gardens at Monticello, 94

Kanst, Friedrich, 55
and South Park, Chicago, 55
Kelly, St. Paul Mayor Randy, 104, 118
Kelper, Adolf, 70
King Construction Company, 19, 64
Knuth, Paul, 113, **115**

Lake Como, 13-16, **14**, **28**, 34, 36, 38, 39, 44, 58-59
Lake (Como) Drive, 39
Lake Harriet, Minneapolis, 35
Lake Phalen, St. Paul, 34, 38
Lantern Lighting Festival, **99**, **100**
Latimer, Alfred J., 35
Lefebvre and Deslauriers, greenhouse builders, 59
Lick, James, 53
Lily Pond, 23, 44, 62
Lincoln Park, Chicago, 33, 52, **53**
Lindley, John, 61
Longwood Gardens, Pennsylvania, 62
and Pierre du Pont, 113-14
Lord & Burnham, 52, 53, 54
Loring, Charles, 37
Loring Park, 22, **37**
and Ole Bull, 22
Loudon, John Claudius, 50
Lucker, Dean, 27

Mannheimer Memorial, 20, **21**
Manship, Paul, 26-27
and *Indian Hunter and His Dog*, **26**, 27
Marshall, William R., 34
Marjorie McNeely Conservatory, 5, 7, 8, 10-11, 19, 38, **81**, **104**, 104, **107**, 107-08, 113-14, 118
Marjorie McNeely Conservatory Bonsai Collection, 107
and Mas Imazumi, 107
and John Naka, 107
Martel, Pierre, 31
Matsuda, Masami, 98-101
McDonough Recreation Center, 92
McKenty, Henry, 13, 14, 19, 36
McKnight Formal Garden, **26**, 27
McKnight, William L. and Maude, 27
McNeely, Donald, 103-04
McNeely, Greg, 104
McNeely, Marjorie, **103**, 103-04
and Guthrie Theater, 103
and Friends of the Minneapolis Institute of Arts, 103
and Minnesota Historical Society, 103
and Radio Talking Book Network, Minnesota State Services for the Blind, 103
Men's Garden Club, 25, 78, 88
Merrill Corporation, 92
and Music Under Glass 92

Metropolitan Council, 89
Minneapolis City Council, 35
Minnesota Bonsai Society, 88
Minnesota Flower Show, 70
Minnesota Horticultural Society, 79, 88
Minnesota State Fair, 36, 109
Minnesota Zoo, 91
Missouri Botanical Garden, 71
Monet, Claude, and Giverny, 90
mosaiculture, 46, 55
Moses, Robert, 42
Mount Vernon, 49
Murphy, Joan, **115**
Murray, William Pitt, 33

Nason, George L., 23, 72
National Mall, 30
National Register of Historic Places, 85
Nelumbium Pond and Rockery, 62
New York Botanical Garden, 52, **54**, 86
Nicollet Island, 35
North Garden, 24, 72, **78**, **90**, 90, **92**, 94, 108, 111
Nussbaumer, Frederick, 38-39, 41, 43-46, **47**, 52, **55**, 55, 57, 59-60, 62, 64, 67, 69, 70, 72, 113-14

Olbrich Gardens, 113
Olmsted, Frederick Law, 30, **32**, 33, 41
and John Singer Sargent, **32**
Orchid House, 107-08, **109**, **110**
Orchid Society of Minnesota, 88, 109
Ordway, Charlotte, 99

Paxton, Joseph, 17, 50-52, 61
and Crystal Palace, **51**
Palm Dome, **25**, 25, 59, **71**, 72, **78**, 85, 89-91, **91**, 108, 111, 116
Patsche, Dave, **68**, 113
Pesek, Bill, 88, 100-01
Phipps, Henry, 52, 54
and Phipps Conservatory, 52, **54**
Pine Grove Walk, 114
Pusey, Pennock, 35

Queen Victoria, 50, 61

Ralph Burke and Associates, Architects, 89
Ramsey, Minnesota Territorial Governor Alexander, 13
Ramsey County, 16, 36
Ramsey County Poor Farm, 36
Ramsey Hill, 26-27
Rice, Henry M., 15
Rice Park, St. Paul, 16
Royal Botanic Gardens at Kew, 41, 48, **49**, 49, 61, 65
and Palm House, **49**, 65

St. Paul and Pacific Line, 14
St. Paul Board of Park Commissioners, 38, **55**
St. Paul Chamber of Commerce, 33, 35
St. Paul Common Council, 16
St. Paul Dispatch, 35
St. Paul Division of Parks and Recreation, 89
St. Paul Garden Club, 99, 103
St. Paul Horticultural Society, 70
St. Paul-Nagasaki Sister City Committee, 78, 98-99, 114
St. Paul Parks and Recreation Department, 38
St. Paul Pioneer Press, 57, 70, 79
St. Paul Winter Carnival, 75, **76**, 79
and Winter Carnival Orchid Show, 108

and King Boreas, 76
and Boreas Rex VIII, 76
and Eva Wicker, Queen of the Snows, 79
Sandy Lake, 13
Schenley Park, Pennsylvania, 52, **54**
Schiffman, Dr. Rudolph, 21, 98
and Schiffman Fountain, **12**
and Louisiana Purchase Exposition, 98
Schiller, Christoph Friedrich von, 22
and Schiller Monument, **22**
Schwietz, Robert, 79-80, 98-99
Shepard, Donald, 24
and *St. Francis of Assisi*, 24
Sibley, Henry H., 33
Siegel, Victor, 55
and Columbia Gardens, Butte, Montana, 55
Sladky, Roberta, 91-92, 94, 104, 113
Smith, Chester R., 64
Smith, Robert, 16
Smith Park, St. Paul, 16
Spring Flower Show, **77**
Stanke, Terry, volunteer, **93**, 93
Stowell, Doe Hauser, 27
Stowell, James, 27
Sunken Garden, 8, 25, **27**, **68**, 72-76, **73**, 78, **79**, 80, 90-91, 94, **95**, 113

Tanguchi, Mikio, 98
Theodore Wirth Park, Minneapolis, 52
Thoreau, Henry David, 29
Thorvaldsen, Bertel, 23
and *Night and Morning*, 23
Toltz Engineering Company, 64
Tropical Encounters, 39, 97, 106-08, 113
Turner, Richard, 49

University of Minnesota, 33
U.S. Botanical Gardens, 94

Vaux, Calvert, 30
Verret, Paul, and Saint Paul Foundation, 94-96
Versailles, 45, 100
Visitor Center, 11, 39, 62, 96, **97**, 97, **102**, 104, **105**, 106-07, 113, **117**

Ward, Dr. Nathaniel, 49
Washington, George, 49-50
Washington, Martha, 93
Wheelock, Joseph A., 33,57
Wheelock School, children from, **78**
Whitney, Cornelius S., 16
Willow Walk, **43**, 43
Winsor/Faricy, 89
Works Progress Administration, 85
World Monuments Fund, 87

Xcel Energy, 110

Yellowstone National Park, 33
Yosemite, 33

Zierden, Ray, 99
and Hamm Memorial Waterfall, 99
Zobota Café, 106
Zonta, 101
Zoological Building, 39, 97

This book was designed with care by

Mary Susan Oleson
NASHVILLE, TENNESSEE